Hawker Sea Fury

in action

By Ron Mackay
Color By Don Greer
Illustrated by Perry Manley

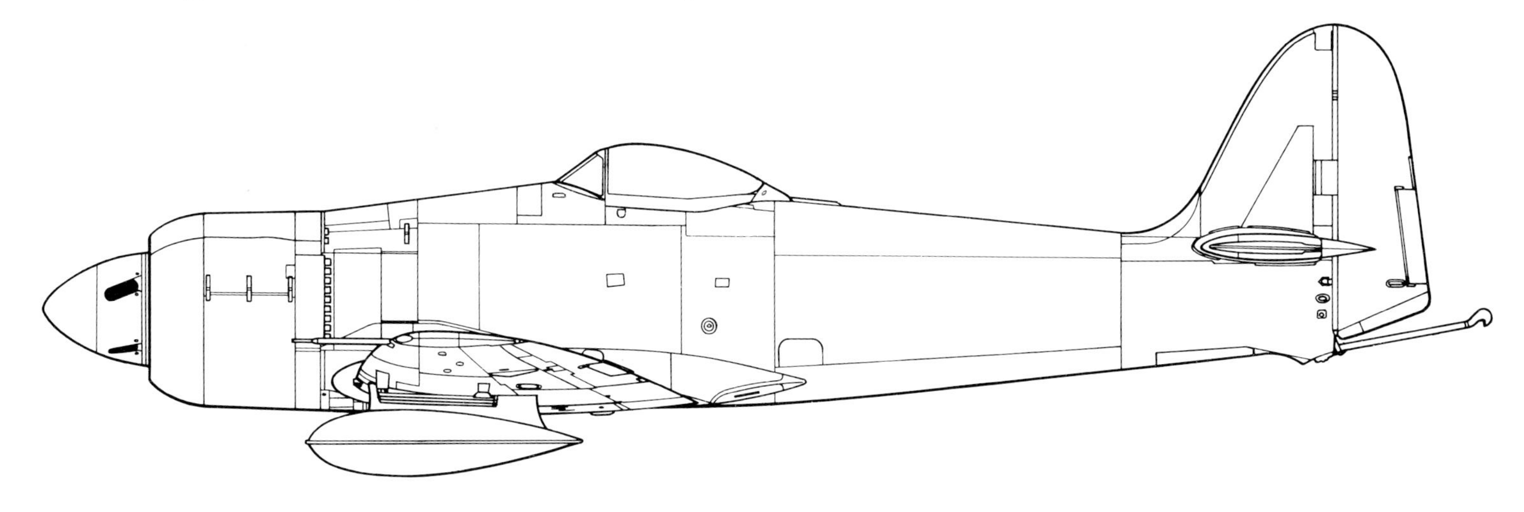

Aircraft Number 117

squadron/signal publications

On 9 August 1952, a flight of Sea Furies from HMS OCEAN was attacked by North Korean MiG-15s. During the fight, a MiG-15 pulled up in front of the flight and was fired on by several members. One MiG-15 was shot down and another damaged, with credit for the destroyed MiG going to LT Carmichael, the Flight Leader.

ISBN 0-89747-267-5

If you have any photographs of the aircraft, armor, soldiers or ships of any nation, particularly wartime snapshots, why not share them with us and help make Squadron/Signal's books all the more interesting and complete in the future. Any photograph sent to us will be copied and the original returned. The donor will be fully credited for any photos used. Please send them to:

Squadron/Signal Publications, Inc.
1115 Crowley Drive.
Carrollton, TX 75011-5010.

Photo Credits

FAA Museum
Royal Navy
British Aerospace
B. J. Lowe
Imperial War Museum
Canadian National Archives
R. C. Ward
Gaston Barnell
Nicholas J. Waters III
J. W. R. Rawlings

This Sea Fury FB 11 carries the CW tail code that identifies it as being stationed at Royal Naval Air Station (RNAS) Culdrose. The aircraft was part of the fourth production batch and carried the standard Royal Navy camouflage of Dark Sea Gray uppersurfaces with Sky sides and undersurfaces.

CW
136
ROYAL NAVY
VX639

Introduction

The Hawker Sea Fury belongs to a select group of piston-engined fighter aircraft whose general performance reflected the ultimate from this form of fighter aircraft power plant.

It is an irony that the Royal Navy became the primary operator of the aircraft, since it was originally designed for the Royal Air Force as part of the succession of fighter aircraft from Hawker that started with the Fury biplane and included the Hurricane, Typhoon and Tempest.

During 1942, Hawker, at the request of the Air Ministry, began work on a long-range fighter for operations in the Far East. When Japan had struck British possessions in the Far East in December of 1941, the only fighters in the area were short-ranged defensive fighters. The distinct possibility of having to drive the Japanese thousands of miles back to their homeland meant that fighter escorts for Allied bomber forces would be faced with the need to fly prolonged distances over water to strike at enemy targets.

Although a variant of the Tempest was being tested with an air-cooled Centaurus radial engine, the design's relatively high wing-loading, compared to Japanese light weight fighters, led many to believe that it was unsuited to the long-range escort role. In the event, such reservations were over-stated since U.S. Navy fighters of similar weights overcame the problem by avoiding dogfights. They simply stood off and used their massive firepower to blast their light weight opponents. This lesson in survival and success, however, was still to be proven during 1942.

The long-range fighter requirement led the Air Ministry to issue Specification F6/42, for what was titled the "Tempest Light Fighter (Centaurus)." If the RAF requirement was deemed urgent, then the needs of the Royal Navy for a similar type was paramount, since the Royal Navy (RN) lacked any sort of long-range carrier fighter. Its principal fighters were either Hurricane or Spitfire adaptations or hopelessly obsolete Naval designs such as the Fairey Fulmar; all were extremely short-ranged.

In the event, Specification F6/42 would prove to be of no practical benefit to the Royal Navy during the Second World War. Indeed, the RN would complete the conflict with its main offensive fighters being American designs (Grumman F6F Hellcats and Chance-Vought F4U Corsairs).

Joint RN and RAF involvement in the Hawker project was officially recognized with the issue of Order F2/43 (RAF) and F7/43 (RN) in January and February of 1943 respectively. Overall production was allocated to Hawker but construction of the intended Naval batch of 200 was assigned to Boulton Paul's Wolverhampton factory. A similar number of aircraft was to be turned out for the RAF with Hawker.

The naval conversion included folding wings, an arresting hook and catapult attachments. It was a year after the original Order F7/43 was issued that the conversion order, Order N22/44, was issued to cover these modifications. In the interim period a separate contract No 26430/43 was awarded for conversion of six airframes to be used for engine and structural tests. Two would be fitted with Centaurus XXII radial engines, two with Rolls Royce Griffin in-line engines and the fifth with a Centaurus XII radial engine. The sixth airframe was used for structural tests.

Although the basic fuselage outline of the Fury/Sea Fury, as the respective RAF and RN variants were named during 1945, bore a strong resemblance to the Tempest there was noticeable reduction in the wing span by some two feet seven inches. Even with a reduced wing span, the aircraft did not have an increased wing loading thanks mainly to a reduced overall structural weight.

Fuel capacity was of prime importance in the development of the Fury/Sea Fury and in this respect the large fuselage size of the new Hawker design was a positive factor, housing no less than five internal tanks. The largest (97 gallons) was located, along with a thirty gallon auxiliary tank, just behind the engine's fireproof bulkhead. There were three other tanks located in the wings, with two positioned between the wing-spars and a third located in a forward section of the starboard wing. Their contents were transferred to the main tank by air pressure from the exhaust side of the engine vacuum pump. A single oil tank containing fourteen gallons was positioned in the engine bay, forward of the firewall. It was equipped with a negative "g" valve to ensure continuous oil flow under all flight conditions. Total internal fuel capacity was 200 gallons with a further 90 or 180 gallons available in underwing drop tanks (45 or 90 gallon).

Of the four RAF assigned airframes, NX798 was the first to fly on 1 September 1944, being closely followed by LA610 on 27 November 1944. Test flights with the former were bedevilled by engine crankshaft lubrication failures which brought about a number of forced landings and at least one belly landing. LA610, fitted with a Griffin in-line engine, was more fortunate; however, the Griffin was later replaced with a Sabre VII, a power plant which had already proved itself on the Hawker Tempest. The Centaurus XVIII, which would power production aircraft, was substituted on the second prototype (NX802). This specific airframe was purchased by Hawker for general development work and was originally equipped with the Centaurus XII. Finally, the fourth prototype (VP207) featured a wide-tolerance Sabre VII, but tests showed no improvement and further development was abandoned.

The first naval prototype was SR661, which lifted off on 21 February 1945. It was a "non-navalised" aircraft having a fixed wing, although it did carry a short-length tail hook. Originally the prototype carried a four blade propeller, which was subsequently replaced by a five blade propeller which would become standard on the production Sea Fury.

The short fin and rudder was found to create a dangerous tendency for the aircraft to swing during take off. As a result, the rudder was lengthened which also cured the problem of a loss of directional stability whenever the aircraft executed an over-shoot. The rigid engine mounts led to excessive engine vibration which affected engine performance. Any sudden application of power, such as in an over-shoot situation, tended to cause the engine to over-speed. Slow throttle response was also experienced in other stages of approach or pulling away, the last thing a pilot needed to face with when operating from a carrier flight deck! With all these problems, it was hardly surprising that the trials conducted during May of 1945 by the Aircraft and Armament Experimental Establishment (A & A E E) were confined to airfields ashore.

The original Fury F2/43 prototype parked on the grass at Langley airfield. The aircraft carries late war camouflage with a Yellow letter P within a Yellow circle on the fuselage. This marking identified the aircraft as a service prototype.

It was during this period that the entire RAF order and half the Royal Navy order was abruptly cancelled. This action was taken because of the end of the European conflict, as well as the impending end of the Pacific War. Also by mid-1945, the need for a long-range fighter in the Pacific was largely nullified by the presence of Allied forces on islands from which available American fighters, especially the P-51, could range over the enemy's heartland.

The RAF was bringing the Meteor and Vampire jet-propelled fighters into service, and the Royal Navy was also turning its thoughts toward the use of jet fighters aboard aircraft carriers. Deck-landing tests were conducted with a modified Vampire during late 1945 and the Supermarine company was developing the T932 jet for the Navy (which would evolve into the Attacker). These events would seem to argue against the continued development of the Sea Fury; however there was a degree of Navy caution in the development of jets that led to the introduction, during the 1950s, of other propeller-driven designs such as the Gannet and Wyvern.

Deck-landing trials were assigned to Sea Fury prototype SR666. This aircraft had a lengthened rudder, five blade propeller, longer tail hook and a tailwheel lock. The lengthened tail hook ended the aircraft's tendency to slew to one side as the arresting wire was engaged, as was common with the original short hook. The tailwheel lock provided a three-fold benefit. First, the aircraft was kept on a straight line during takeoff and landing. Second, it countered the problem of directional instability with the single point aircraft towing hook (this tow hook also served as the catapult attachment point). Third, the automatic fore and aft alignment of the wheel allowed it to retract properly into its well. Locking also meant that the wheel was better able to withstand the acceleration during launch, unlike the non-locking unit tested on SR661 at RAE Farnborough which collapsed during the first simulated launch!

SR661 and SR666 had flown with the VII and XV variants of the Centaurus engines. The third Sea Fury prototype (VB857) was powered by a Centaurus XXII in place of the intended XVIII and made its initial flight on 31 January 1946. Of prime importance to the success of the design was the dynafocal base engine mounts which had the effect of almost totally eliminating the vibration suffered by the earlier prototypes. This in turn eliminated the dangerous throttle-lag and engine over-speeding. There remained one engine related problem, however — carbon monoxide leakage through the firewall into the cockpit. This proved difficult to counter and it was ordered that pilots would wear oxygen masks at all times. VB857 was a fully "navalised" prototype with folding wings and a five blade propeller.

SR661 was the first Sea Fury navalised N7/43 prototype. The aircraft carried RAF style camouflage, was fitted with a short tail hook and the wings did not fold.

The second Sea Fury prototype (SR666) was fitted with folding wings and armament. It was powered by a Centaurus XV air cooled radial engine and had a five blade propeller installed. The aircraft took part in pre-production deck trials on board HMS VICTORIOUS during 1946.

Development

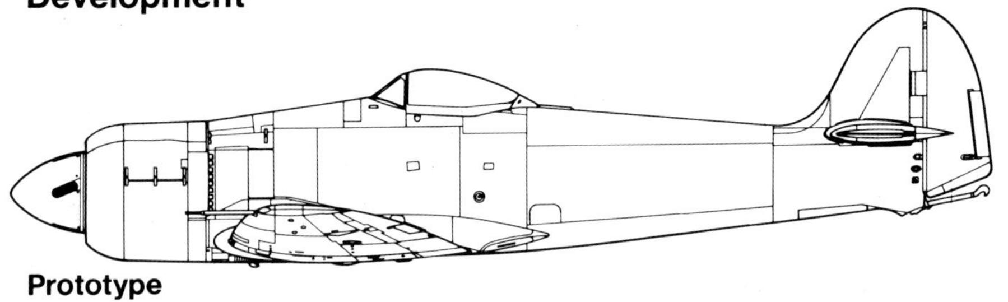
Prototype

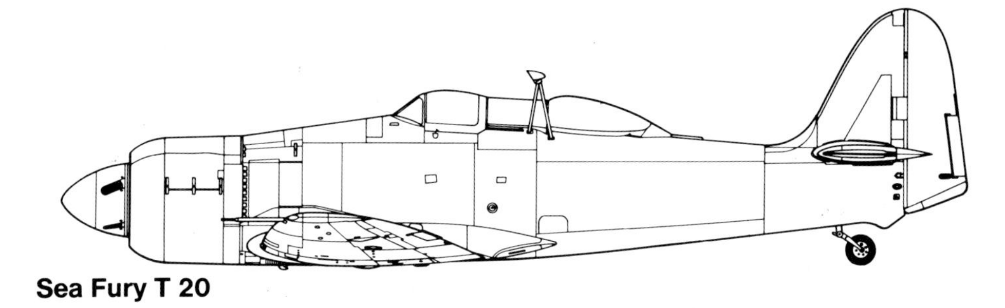
Sea Fury T 20

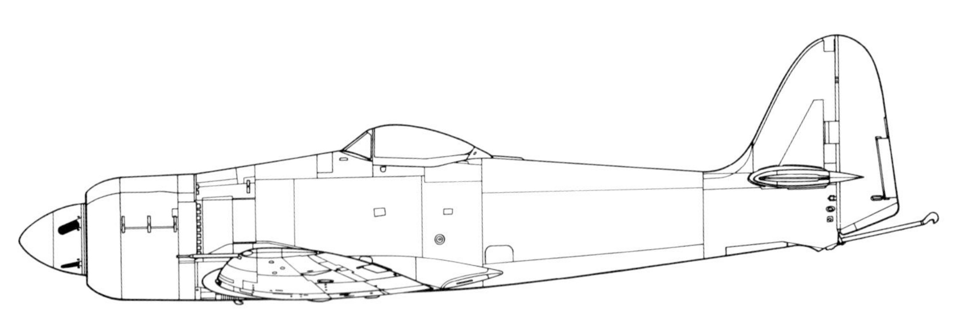
Sea Fury Mk X

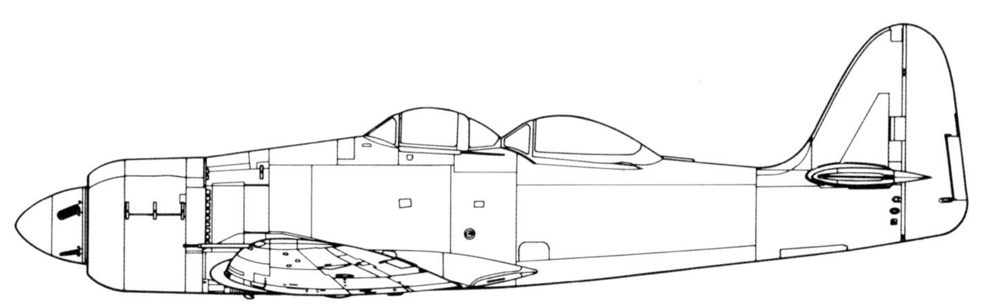
Sea Fury T 61

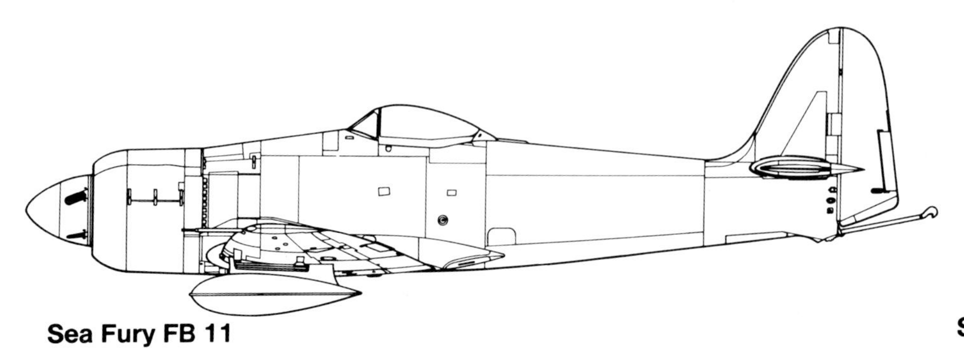
Sea Fury FB 11

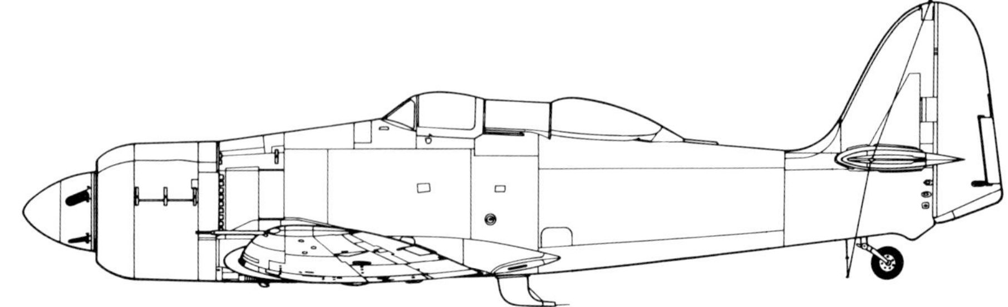
Sea Fury TT 20

Sea Fury Mk X

The first production Sea Fury (TF895), along with another forty-nine aircraft, was given the designation Sea Fury Mk X. These aircraft were powered by a 2,550 hp 18 cylinder air-cooled Bristol Centaurus engine driving a four blade propeller. The Sea Fury Mk X was intended to serve in the air superiority fighter role armed with four Hispano 20MM cannons with a total ammunition capacity of 580 rounds.

Most of the first twenty aircraft produced were diverted either to Hawker or Bristol for use in a variety of experimental roles. TF895 and TF897 were used for manufacturer's trials, while TF902 and TF908 went to the A & AEE for armament tests. Some of the manufacturer's experiments involved spring-tabs which were torsion-bar operated and fitted to both the ailerons and rudder. These gave the aircraft a high rate of roll, which was a great advantage in combat. This modification was later introduced on the production line and retrofitted to earlier aircraft.

Service clearance took nearly twelve months and during that period armament tests at Boscombe Down (including trials with underwing bomb loads), as well as more deck-handling trials were carried out. Flight tests with the five blade propeller, fitted to the third Sea Fury prototype (VB857), resulted in this propeller becoming standard equipment on all Sea Furies and being retrofitted to earlier aircraft.

Carrier compatibility trials were completed by early 1947 and Service Clearance was granted that same Summer. Operational training using three Mk X aircraft assigned to No 778 Squadron was marred by the crash of TF906, while the sole aircraft allotted to No 807 Squadron, the first unit scheduled to re-equip with the Sea Fury, was also written off in a crash. In the event, these accidents had no effect on the overall re-equipment plan and between August of 1947 and February of 1948, Nos 802, 803 and 805 Squadrons joined No 807 in converting to the Sea Fury.

This was the fourth Sea Fury Mk X off the production line. The first ten aircraft off the production line were equipped with four blade propellers, but were later retrofitted with the five blade propeller. The Mk X differed from the first prototype in having a larger rudder and lengthened tail hook.

The Centaurus XVIII air-cooled radial engine provided 2,480 hp on takeoff, 2,550 hp at 4,000 feet and 1,600 hp at 21,000 feet. This gave the Sea Fury Mk X a maximum speed of 465 mph at 18,000 feet, a rate of climb of some 3,000 feet per minute, a service ceiling of 36,000 feet, and a range of 710 miles (with the Rotol five blade propeller).

The original intention for the Sea Fury to function as an air superiority fighter was soon changed. During February of 1948 the Seafire Mk 47 fighter was introduced into service and this event caused Naval authorities to question the need for two separate designs to fill the same role. The Sea Fury's solid and rugged airframe lent itself naturally for use as a fighter-bomber and the aircraft was already cleared to carry underwing bombs. As a result, Hawker was requested to convert the second and subsequent production batches to the fighter-bomber configuration.

Various weapons trials had been carried out with SR666 and TF923. During these tests, bombs, rockets, mines, containers and RATOG (Rocket assisted Take-off Gear) was installed and successfully tested. Information gained during these tests helped Hawker reconfigure the Sea Fury Mk X for the fighter-bomber role with a minimum of changes.

Tail Development

First Sea Fury Prototype

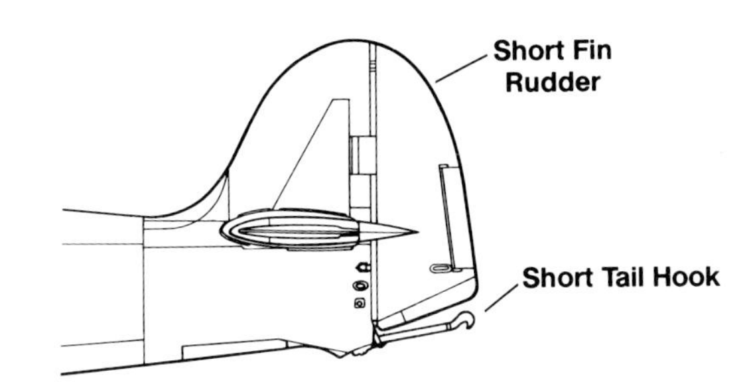

Sea Fury Mk X

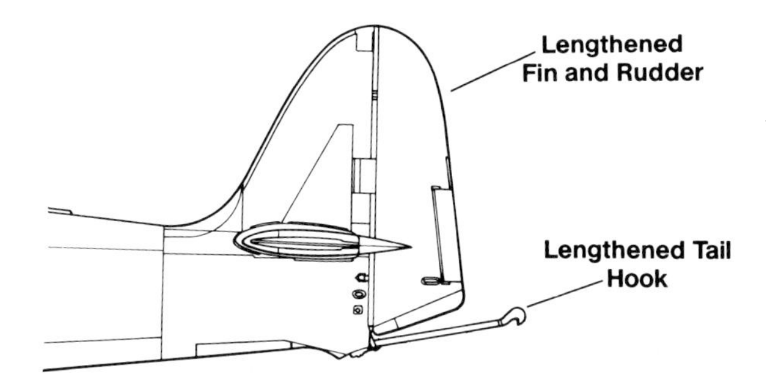

A Sea Fury Mk X of No 805 Squadron over Portsmouth harbor during 1948. The aircraft was the seventh Mk X built and has been retrofitted with a five blade Rotol propeller. The aircraft carries the early Sea Fury camouflage of Extra Dark Sea Gray uppersurfaces over Sky undersurfaces.

An early Sea Fury Mk X chocked on the forward elevator of HMS VICTORIOUS. The aircraft has been retrofitted with a five blade Rotol propeller in place of the original four blade prop. The small opening in the port wing root was for the engine oil cooler.

Propeller

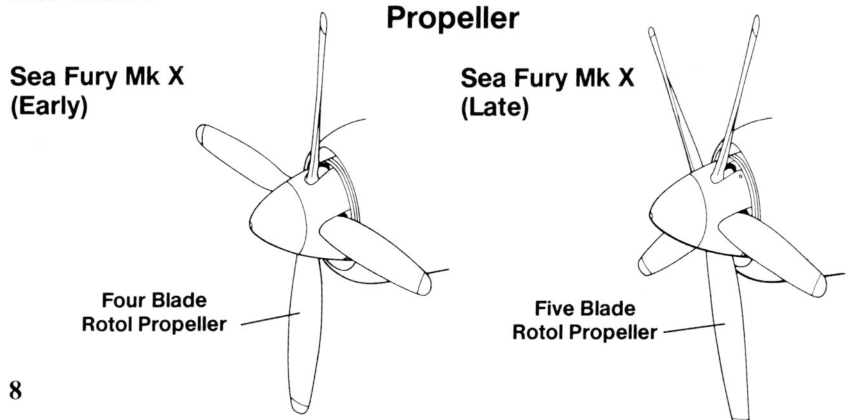

A flight of four Sea Fury Mk Xs of No 802 Squadron flies over Capetown, South Africa during 1948. Sea Fury camouflage was undergoing a transition during this period and the third aircraft carries the new Sea Fury camouflage scheme.

Even though it was intended to be a pure fighter, the Mk X could carry two bombs. This capability led to the later conversion of the Sea Fury to the fighter bomber role. The cannons were slightly staggered with the outboard gun being slightly further forward than the inboard guns.

With its flaps fully deployed, an early production Sea Fury Mk X (TF898) receives the "cut power" signal from the Deck Landing Officer aboard HMS VICTORIOUS. The Sea Fury Mk X was intended as a fleet defense air superiority fighter.

Sea Fury FB 11

Based on the Royal Navy's decision to convert the remaining Sea Furies under contract to the fighter bomber configuration, the fifty-first Sea Fury airframe on Hawker's production line received a new designation — Sea Fury FB 11. (This aircraft was TF956, which would serve in the Fleet Air Arm (FAA) for some fifteen years prior to being repurchased by the Hawker company for restoration and inclusion in the Royal Naval Historic Flight).

Externally, the Sea Fury FB 11 differed from the Mk X in having underwing attachment points for bombs, drop tanks and rockets. Internally, there were some fifty minor changes. The overall performance of the Sea Fury FB 11 was close to that of the earlier Sea Fury Mk X with the exception of rate of climb. With an increase in loaded weight of 1,840 pounds (10,660 pounds to 12,500 pounds) the time to 30,000 feet increased to 10.0 minutes, compared to 9.8 minutes for the Mk X. With the addition of underwing drop tanks range was increased from 700 to 1,040 miles.

The internal armament of four Hispano Mk 5 20MM cannons was retained, while external underwing offensive loads could now include either two 500 or 1,000 pound bombs on underwing pylons mounted just outboard of the main landing gear, twelve 3 inch rockets (three double launchers per wing) or four 180 pound Triplex rockets.

The Sea Fury FB 11 became the primary production variant with initial deliveries to FAA units commencing during May of 1948. The first squadron to receive the FB 11 was No 802 Squadron followed closely by Nos 801, 803, 804, 805, 807 and 808 Squadrons. Three of these squadrons were assigned to Carrier Air Groups: 801 to 1 CAG, 807 to 17 CAG and 804 to 21 CAG.

Over the course of Sea Fury production, the Royal Navy took delivery of a total of 615 Sea Furies, the majority of these being FB 11s.

This Sea Fury FB 11 of 806 Squadron crashed, flipped over and exploded aboard HMS ILLUSTRIOUS on 25 May 1949. The Seaman in the foreground was one of two who later received awards for bravery after pulling the pilot clear of the crash.

This Sea Fury FB 11 (WF619) was part of the fourth FB 11 production batch, produced under Contract 2576/5/7/48. The aircraft was camouflaged in Dark Sea Gray uppersurfaces with Sky sides and undersurfaces. The national markings were restricted to wing and fuselage roundels only with no fin flash being carried.

Still carrying the early style Sea Fury camouflage, this FB 11 (VW238) was participating in Arctic trials on HMS VENGEANCE. The wing roundels are located on the wing tips due to the method of serial number placement. The aircraft has the rocket rail supports in position, with no rails in place between them.

A Sea Fury FB 11 rolls over after engaging the carrier's crash barrier. This Sea Fury was assigned to the Aircraft and Armaments Experimental Establishment (A & AEE) but this trial flight was almost certainly its last.

When HMS WARRIOR completed her Far East tour, she called at Port Elizabeth, South Africa for an open house. These civilian visitors admiring Sea Fury FB 11 WE 725 were some of the 10,000 visitors that toured the ship during a single afternoon. The Firefly in the background carries a radar pod on its port wing.

Seven Sea Fury FB 11s of No 898 Squadron use their engines to assist the carrier in maneuvering within the narrow confines of Malta's grand Harbor. This was known as "pin-wheeling" and was hard on aircraft engines.

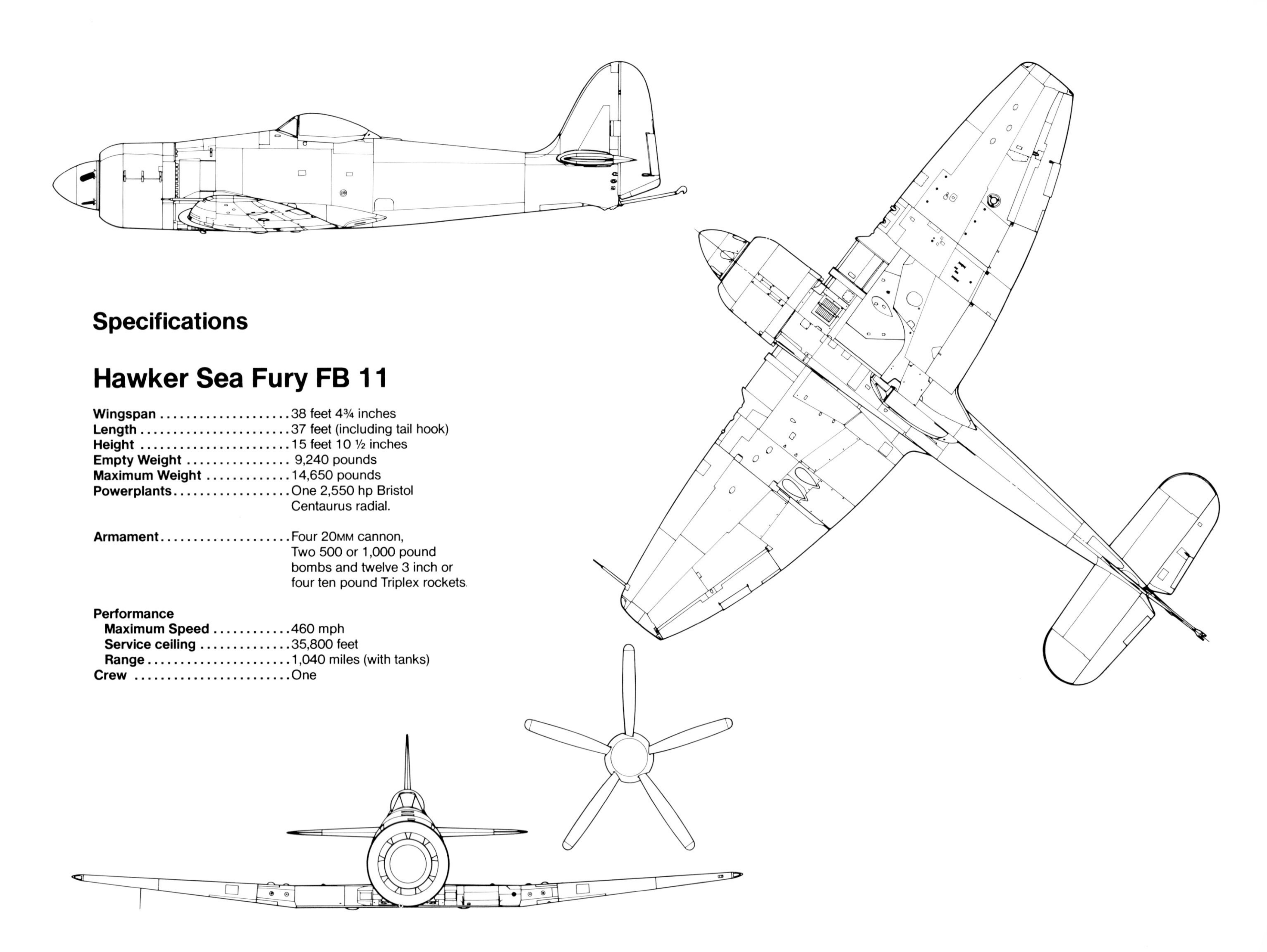

Specifications

Hawker Sea Fury FB 11

Wingspan	38 feet 4¾ inches
Length	37 feet (including tail hook)
Height	15 feet 10 ½ inches
Empty Weight	9,240 pounds
Maximum Weight	14,650 pounds
Powerplants	One 2,550 hp Bristol Centaurus radial.
Armament	Four 20MM cannon, Two 500 or 1,000 pound bombs and twelve 3 inch or four ten pound Triplex rockets.
Performance	
Maximum Speed	460 mph
Service ceiling	35,800 feet
Range	1,040 miles (with tanks)
Crew	One

Korea

Just five years after the Second World War, America and Great Britain were involved in yet another major war, although in the diplomatic words of the United Nations it was termed a "Police Action." Thanks to the fortunate absence of Russia from the UN Security Council, the vote to assist South Korea in countering the brutal invasion by North Korea passed without a Soviet veto. As a result of this vote, combat forces, under UN sponsorship, were dispatched to Korea.

Air power was to be quickly and comprehensively employed by the UN Forces and the United States in particular. During the fighting, both propeller and jet-powered aircraft would prove their full value to the allied side. In the case of Great Britain, it was mainly represented by the Royal Navy and Fleet Air Arm. The Royal Navy deployed Light Fleet Carriers off Korea to support UN operations. These ships carried the Firefly and the Sea Fury FB 11.

Although the Navy did have its first jet fighter (Supermarine Attacker) in service by 1951 and was just bringing its second jet fighter (Hawker Sea Hawk) into service, it was to fight the Korean War with propeller-driven aircraft. At first it seems strange that the Royal Navy did not commit the Attacker (a rough equivalent of the American F2H Banshee and F9F Panther) but instead depended on aircraft which were out-performed in the air by the MiG-15.

Most of the air-to-ground tactical air strikes in Korea were against targets tucked away in hilly terrain. Many of these targets were of the type that required frequent re-strike and were defended by strong anti-aircraft defenses. In this type of war the more sturdily built piston-engined fighter-bombers could and did absorb a far greater degree of battle damage than could their jet-powered contemporaries.

In the case of the Sea Fury, the better tactical handling of the aircraft by their pilots meant that losses were kept to a minimum even on the few occasions when MiG-15s were encountered and in some cases the MiGs came out on the losing end. Nevertheless, it is unlikely that the crews' loyalty to the Sea Fury would not have prevented them from wishing they had the greater power and speed of a jet on such occasions!

In addition to the Royal Navy carriers HMS TRIUMPH, THESEUS, GLORY and OCEAN, Australia sent the HMAS SYNDEY. Additionally, HMS UNICORN acted as a ferry bringing replacement aircraft out from Hong Kong. Carrier operations off Korea were difficult: steam catapults, angled decks and landing-mirror systems were as yet non-existent and landing on a snow-covered deck with a mass of parked aircraft ahead had to be a first pass success since a wave-off was virtually impossible. The sturdy undercarriage of the Sea Fury at least allowed for the aircraft to be "dropped" (full stall landings) onto the deck with little risk of a landing gear collapse. Maintenance duties were never easy especially on the open deck in the forbidding weather conditions that often existed in this part of the World.

The carriers were "on station" for eleven days. Usually there were four days of operations with three days for replenishment followed by four days of operations. After this line period, the ship was relieved by a U.S. Navy carrier and proceeded to Iwakuni, Japan, for a "rest and relief" period. Although short in terms of overall time, on each operational line period each aircraft carried out between two and three sorties a day.

Although HMS TRIUMPH was the first Royal Navy carrier to participate in the Korean War, it was on HMS THESEUS that the first Sea Fury operations took place with No 807 Squadron. The carrier was part of the West Coast element of Task Force 95 whose primary duty was to act as a blockade force. The carrier and four escort destroyers comprised TF95.11. This force was later joined by the light carrier USS BATAAN with her escorts.

Apart from keeping up a blockade, carrier duties included the prevention of enemy amphibious landings, the protection of sea convoys and air support of UN ground forces. TF 95.11 conducted daily reconnaissance of the West coast to check on enemy shipping movements, mine-laying attempts and to monitor the special fishing sanctuaries created for South Korean vessels.

Combat Air Patrols (CAP) and Anti-Submarine (AS) patrols were a regular feature of Korean operations. In the anti-submarine role at least one Firefly was sent up with the ASH radar set replaced by a 55 gallon nacelle fuel tank. Although this meant that surveillance was totally visual, the extra endurance the tank gave was regarded as worthwhile. In the event, enemy submarine activity was almost totally absent during the entire conflict.

CAPs initially consisted of four Sea Furies but this was later cut to one section (two aircraft). There was a concern that Chinese or Russian aircraft might launch strikes against the carriers from the Shantong Peninsula, some 150 miles west of the carrier's normal station. As the war progressed the reduced number of fighters forming the CAP was felt to be sufficient to meet the threat. There were never any actual combat interceptions by CAP aircraft, except for "Bogies" not displaying IFF. These totaled some 100 aircraft during the war, mostly lost USAF B-29s.

Armed reconnaissance sorties involved flights well behind enemy lines to hit at specific targets. These targets might be defended by a few troops with rifles or by a well formed and emplaced AA system. This duty was tied in with close air support to the UN ground troops. Both duties were coordinated via the Joint Operations Center (JOC) initially based in Seoul and later (from January of 1951) in Taegu. The JOC was also the link between the American 8th Army and the 5th Air Force. Three US and a UN liaison officer manned the post. At the start of each operational period, each carrier sent a courier aircraft to collect the Special intelligence material and target assignments from JOC.

A mix of Sea Furies and Fireflies tied down on the deck of HMS GLORY as the ships rolls in heavy seas off the Korean coast. The harsh weather conditions, particularly in winter, presented serious problems to safe aircraft carrier operations off Korea.

Deck landings imposed a tremendous strain on the aircraft's landing gear. This Sea Fury FB 11 of No 804 Squadron, serving on HMS GLORY during its first Korean tour, suffered a broken starboard main mount shortly after landing. The Korean identification stripes were Black and White.

A Sea Fury FB 11 is struck below (Royal Navy slang for lowering the aircraft on the carrier's elevator) on HMS GLORY. This Sea Fury was assigned to either No 804 or No 801 Squadron aboard GLORY.

During actual operations, information such as "bomb lines," targets for 5th AF attacks and flak positions were passed by message. For armed reconnaissance sorties special "overlay" traced maps were supplied from the JOC. These maps displayed enemy territory as a series of lettered areas with colors/numbers for the main roads. The armed-recce attacks soon closed the roads to daylight traffic. In the process of these missions much experience was gained in detecting camouflaged target locations, since the North Koreans were very adept at camouflage.

Attack height was usually 1,500 feet but this rose to around 5,000 feet when attacking towns or when approach to the target was likely to encounter heavy defensive zones. Bombing strikes were mainly flown against bridges. Some road bridges proved capable of being bypassed because they were next to built-up fords. Wherever possible, attacks were concentrated upon bridges without such fords. Railway bridges were even more difficult to knock out if the main supports remained intact because they could be quickly "spanned" by Bailey Bridge equipment. Consequently, delayed-action fuses were fitted (6/8 hour periods) and dropped by the aircraft making the last sortie of the day to hinder enemy repair operations.

Close Air Support was a continuation of the WW II principle of backing up ground assaults or helping ward off enemy assaults. To coordinate air support operations, some twelve Tactical Air Command Posts (TACP) were attached to the various army Brigades/Divisions. Each TACP had four Forward Controllers (FC) with one or two AT-6 Texans. The FCs had three/four radio frequencies and acted as aerial fire control directors. After reporting in to a TACP the attacking aircraft were passed on to the joint control of an FC, with the majority of attacks involving bridges or enemy troops.

Bombs were the primary RN weapons, while USAF/USN aircraft used napalm when attacking enemy troops. The effectiveness of Royal Navy attacks were revealed by a member of the Middlesex Regiment: "Your Furies had quite a lot of fun the other day on Hill 1036 in close support of our regiment. We were but 400 yards away from the hill peak when your boys chucked in everything they could lay their hands on. It was beautiful. We were there three minutes after the last plane dived in and saw the mess — it was super. It was a great day for the Navy!"

HMS THESEUS arrived "on station" and conducted operations between 9 and 22 October 1950. Its aircraft launched attacks as far north as Pakchon and Chonju. The second line period (between 29 October and 5 November) coincided with the onset of the Chinese "volunteer" human wave attacks. Despite this threat it was initially decreed that there was no further requirement for carrier operations in the Yellow Sea. This order was soon cancelled and HMS THESEUS arrived back on station during late November. During December some 630 sorties were launched within a seventeen day operational period. Before the ship was relieved in May by HMS GLORY, the air wing flew a total of 3,489 sorties, an average of around 120 sorties for each Firefly and Sea Fury on board.

Weather conditions in the Yellow Sea, for the most part, proved favorable. Flying over Korea in clouds meant the pilots had to be extremely careful due to the mountainous terrain. It was a tribute to their skill that there were no known losses to inadvertent contact with high ground, even though a number of CAS sorties were flown around ridges close to the cloud base. Four Search and Rescue (SAR) helicopters were on hand, and their crews saved four pilots who came down behind enemy lines and a further four who ditched or baled out over the sea. Later a U.S. Navy plane guard helicopter was assigned and its presence meant that aircraft could be operated in higher wind conditions.

While in the Yellow Sea the usual carrier station was about twenty-five miles west of Clifford Island (70/80 miles off the Korean coast). This location was determined by the presence of shallow water areas between the station and the coast. As a result, a USN destroyer was positioned between the carrier and the coast to serve as a SAR vessel.

Planned mission duration was around two and a half hours with external fuel tanks. The added weight of the tanks, however, meant the Sea Furies had to use catapults or RATOG. To take off with two 500 pound bombs, the Sea Fury required at least twenty-eight knots of wind over the bow (even with RATOG). HMS THESEUS had a maximum speed of twenty-two and a half knots (because her keel needed cleaning) and this, along with winds that might drop away with little warning, led to bombing sorties being confined to the Fireflies (which required twenty-one knots over the deck). This meant that the Sea Furies normally made their attacks armed with rockets.

Catapult intervals gradually became so practiced that HMS THESEUS with her single catapult equalled and sometimes surpassed USS BATAAN, even though the BATAAN had two catapults. The desired daily mission rate was fifty sorties and the peak sortie total for any one day was sixty-six. The normal mission launch cycle was two and a half hours.

The two and a half hour launch cycle was adequate for refueling and rearming but there were frequent occasions when a returning aircraft required the barrier be rigged for an emergency landing. This meant that aircraft ready for the next sortie had to be spotted forward. Since the underwing ordnance loads prevented the Sea Fury from folding its wings, it required more time to carry out the respot, which tended to throw out that day's operational mission rate.

A further restriction to Sea Fury ordnance loads was ordered by the Admiralty. A maximum underwing load of four rockets was ordered whenever drop tanks were carried (which with the Sea Fury was on most sorties). Although the poor wind conditions in the Yellow Sea was the main reason, it was just as likely that the available ordnance storage facilities on the Light Fleet Carriers were just sufficient to match this rate of usage, especially with the high rate of sorties being flown.

Squadron pilot strength exceeded aircraft by three pilots and the expected flight hours average for each pilot was set at 40 hours per month. Losses to flak was eight aircraft and four more force-landed at Allied bases. Despite this, morale was sustained by a combination of fairly definite time-limits for on station periods, adequate training periods, as well as the knowledge that there was a good U.S. helicopter rescue facility on hand to support any downed pilots in enemy territory or at sea.

HMS THESEUS was replaced by HMS GLORY with No 804 Squadron. One of their first duties was Operation STRANGLE, aimed at destroying enemy communications within a region extending above the 38th Parallel. Following this, a ground offensive was launched during July which forced the North Korean/Chinese units back over the 38th Parallel. Unfortunately, the enemy counter-attacked and brought the battle-line back to the June 1950 line.

In what would be the first of three separate tours of duty, the air wing on HMS GLORY flew no less than 2,892 sorties and one pilot (LT Young) completed no less than 100. Although very little air combat occurred, there was a problem with aircraft recognition. On 25 June 1951, four Sea Furies were attacked by USAF F-80 Shooting Stars; happily, there were no losses or casualties!

The importance of the presence of U.S. helicopters was highlighted during June when with the GLORY's aircraft was out of service. As a result, GLORY's commander, CAPT Colquhoun, shifted his operations area about thirty miles north of its normal location so that his ship might be as close as possible to land-based rescue units. The pace of operations was always fast and during Operation STRANGLE, no less than eighty-four sorties were launched in one twenty-four hour period.

HMS GLORY was relieved in the Autumn by HMAS SYDNEY with Nos 805 and 808 (RAN) Squadrons. They remained on station until January of 1952. During some forty-three days of available flying weather, they sent out a daily average of fifty-five sorties.

A Sea Fury of 801 Squadron is launched from the catapult on HMS GLORY. The crewman on the left runs across the deck to retrieve the launching bridle to position it for the next aircraft in line. The Sea Fury is armed with three inch rockets.

HMS GLORY was back on station relieving HMAS SYDNEY during March of 1952, for what was to be a relatively short spell ending in April. Along with HMS OCEAN, GLORY would share the Royal Navy carrier commitment for the rest of the war. Her overall sortie total was 1,943 and aircraft losses were some twenty-seven aircraft and nine aircrews. During this time a change was made to Sea Fury armament loads. 500 pound bombs with 30 second delay fuses became the normal ordnance load. It was felt that the bombs could be aimed more precisely with less risk of being caught in the explosion. Also it was thought that the bombs packed a greater "punch." The Sea Fury unit on board HMS GLORY was No 801 Squadron.

The Australian contribution to the Korean War was HMAS SYDNEY and her air group. This Sea Fury FB 11 of No 85 (RAN) Squadron runs its engine up prior to taxiing into position for launch on another sortie. HMAS SYDNEY also operated a second Sea Fury squadron, No 808 (RAN) Squadron.

HMS OCEAN began its first tour during May, which lasted until October of 1952. The general policy was for a few of the veteran crews from the previous carrier to transfer to the relief carrier as advisers. Six ex-GLORY pilots were onboard and one of the first recommendations was for No 802 Squadron pilots to change their angle of attack from 65 to 45 degrees when dive-bombing.

Although the flight decks of Light Fleet Carriers were only marginally shorter than Fleet Carriers, the normal bomb load of two 500 pound bombs meant that there was little margin for safe operations whenever the catapult was out of action. Such circumstances called for the use of RATOG. The jettisonable rocket tubes were located below the fuselage and just behind the wing trailing edge and were fitted in groups of two or three on each side depending on he loaded weight of the aircraft. Prior to takeoff, the pilots would study a chart which showed the RATOG firing point in relation to the actual wind speed over the bow. After launch and when reaching a safe height, at a minimum airspeed of 150 knots, the RATOG Master Switch was set to OFF and the rocket tubes were jettisoned.

HMS OCEAN's commander had made arrangements for the supply of 1,000 pound bombs and any aircraft carrying two of these weapons had to use RATOG. LCDR Shotton, commander of No 802 Squadron, made the first operational launch carrying two 1,000 pound bombs. On launch he forgot to hit the Master Switch and the rockets failed to fire. He staggered off the deck and sank out of sight. Fortunately, he had gained enough momentum to ease his Sea Fury to a safe altitude and continued on with his mission.

Marine ground crewmen inspect this Sea Fury from HMAS SYDNEY which was forced to divert to a U.S. Marine base in Korea. The aircraft in the background is a USMC F4U Corsair.

Doubling of the bomb load meant that the ground attack sorties gained even greater success, but the attackers experienced a steady casualty rate as well. An average of 76 daily sorties was recorded and on one day 123 aircraft were launched. On that day, four breached bridges, a destroyed oil dump and hits on four coastal gun emplacements were recorded. Attacks on sluice gates and night attacks on road transports were flown during September, while attacks on rail bridges were so successful that by the end of HMS OCEAN's line period, CAPT Evans was able to report that all locations on the Pyongyang/Chinnampo rail line had been destroyed.

Enemy jet incursions increased during July and August and Allied top cover was required whenever ground attack operations north of the Taedong River were briefed. The ground attacks were planned to coincide with jet fighter sweeps against MiG Alley. Some enemy fighters, however, broke through the screen and intercepted HMS OCEAN's aircraft. On 2 August a group of eight MiG-15s hit the morning's mixed assault force of Fireflies and Sea Furies. In a sharp but short combat, one Firefly was crippled but the Sea Fury elements inflicted hits on two jets, one of which was seen to crash. This "kill" was credited to the flight leader, LT P. Carmichael. The following day a similar number of jets were met and this time only maneuvering by the Sea Fury pilots and the availability of nearby cloud cover prevented serious casualties, although one MiG-15 was claimed as damaged.

OCEAN had completed ten "on station" patrols by the time she was relieved by HMS GLORY, who in turn completed twenty-five similar patrols within her three operational line periods. GLORY's pilots now concentrated upon cutting rail lines in difficult geographic areas in order to render repair work more difficult. Pre-dawn attacks on road transport continued. HMS OCEAN took over from HMS GLORY during May of 1953 and was to see the war through to its end. The Commonwealth carriers and their Sea

Fury and Firefly squadrons had played a considerable part in the war effort and had helped bring the war to its conclusion.

With the end of the Korean War, the Sea Fury was steadily withdrawn from front line service. During 1953, Nos 801, 804 and 808 Squadrons began converting to the Seahawk jet fighter and No 803 Squadron had already switched to the Attacker F1 during late 1951.

During 1951, the Sea Fury began to replace the Seafire F17s that were operated by several RNVR units. Two reserve squadrons, Nos 1831 and 1836 were the first to receive Sea Furies and by early 1955 six RNVR squadrons were flying the Sea Fury. Their use by RNVR was short, however, and by late 1955, most had been replaced by Attacker and Seahawk jet fighters.

With its flaps and tail hook down, a Sea Fury FB 11 of 807 Squadron comes in for a landing aboard HMS THESEUS during operations off Korea. The aircraft is equipped with bomb pylons outboard of the underwing fuel tanks. The Black and White stripes have covered over the aircraft serial number on the wing undersurface.

"Sammy the Stork" was a temporary visitor aboard HMS GLORY during May of 1952. The bird seems to have taken a liking to this fully loaded Sea Fury FB 11. The bird remained aboard ship for several days and was fed "Kippers."

A Sea Fury from No 801 Squadron, based on HMS GLORY during 1952, went over the side after missing the arresting wires on landing and quickly sank. The fate of the pilot is unknown.

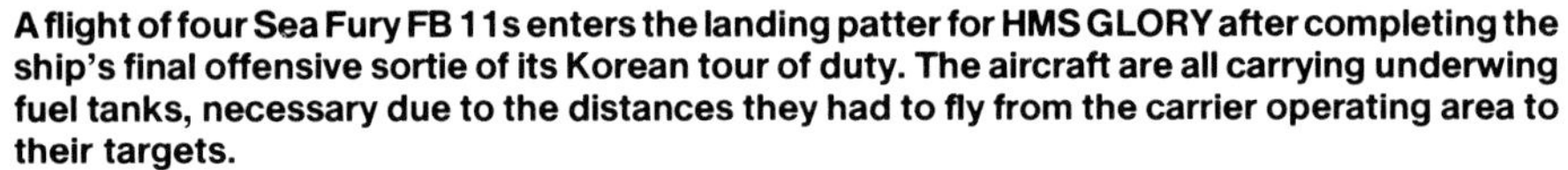

A flight of four Sea Fury FB 11s enters the landing patter for HMS GLORY after completing the ship's final offensive sortie of its Korean tour of duty. The aircraft are all carrying underwing fuel tanks, necessary due to the distances they had to fly from the carrier operating area to their targets.

This Sea Fury FB 11 made a bad approach to HMS GLORY and is attempting to "go around." The ship was on its third and final tour of duty off Korea, between November of 1952 and May of 1953.

Underwing Racks

Rocket Rails

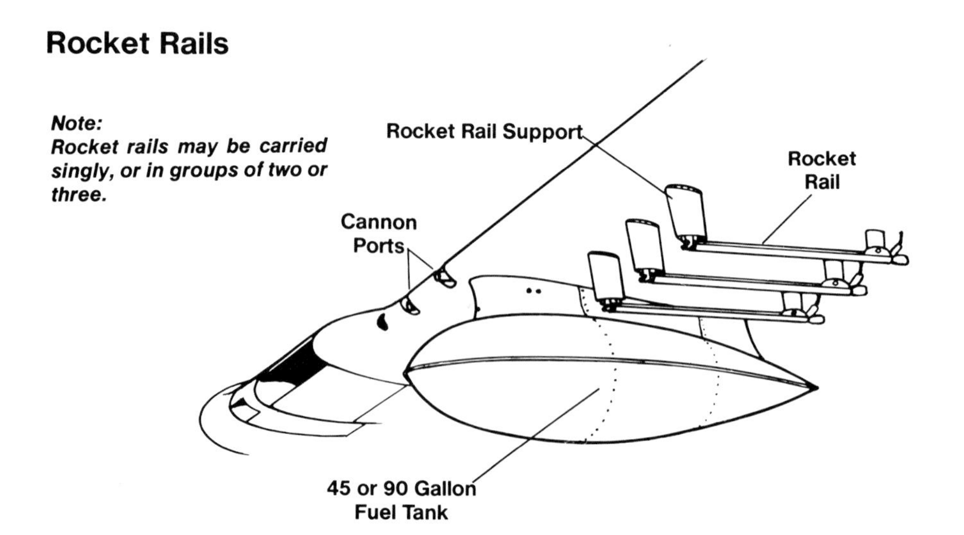

Bomb Rack

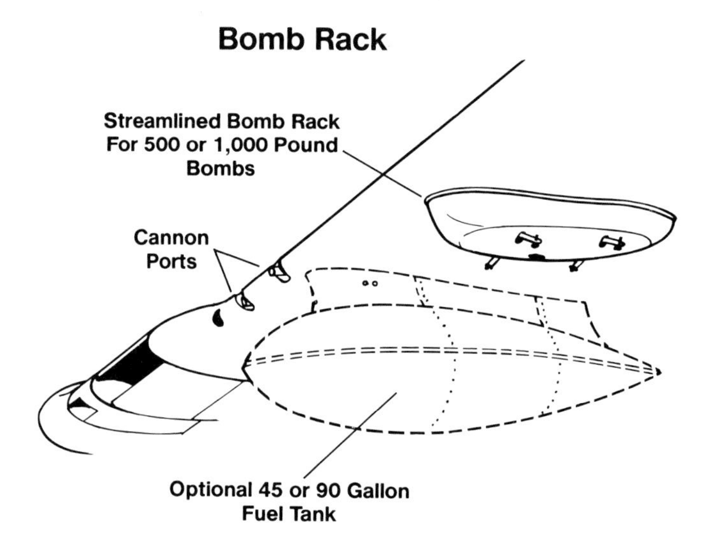

This Sea Fury FB 11 was displayed with the various ordnance it could carry. The underwing rocket rails are configured with double rocket launchers. The aircraft to the left is a Sea Fury T 20 trainer (the 43rd production aircraft).

With a deck load of Sea Furies and Fireflys, HMS THESEUS prepares to depart from the Firth of Forth in Scotland for Exercise MAINBRACE during 1952. The carrier in the background is the Canadian carrier, HMCS MAGNIFICENT. Both vessels are over-shadowed by the massive Forth Bridge which carries mainline railway tracks.

This Sea Fury of No 807 Squadron crashed aboard HMS OCEAN during August of 1953. The aircraft was lowered onto OCEAN's forward elevator and moved to the hangar deck where all usable parts would be salvage from the aircraft if it couldn't be repaired.

A lone Sea Fury is parked on the deck of HMS PERSEUS along with a a group of U.S Navy F9F Panthers, F2H Banshees and F3D Skyknights. These aircraft were all on board to test the Royal Navy's prototype steam catapult. This catapult was later adopted by the U.S. Navy.

This Sea Fury FB 11 of No 1931 (RNVR) Squadron was doing carrier qualification landings on HMS ILLUSTRIOUS during July of 1953 when it hit the arresting wire frame in the foreground and nosed over. The squadron was based at RNAS Stretton.

Carrying "invasion stripes," three Sea Furies prepare for launch from HMS THESEUS. The stripes were a special identification marking used for an exercise involving the Home and Mediterranean Fleets during 1953.

The pilot of this fiercely burning Sea Fury FB 11 of No 801 Squadron desperately scrambles clear as the flight deck crew aboard HMS INDOMITABLE fight the fire with foam extinguishers. The aircraft probably caught its main landing gear in the deck over-hang on landing.

A trio of No 810 Squadron Sea Furies shares the flight deck of HMS CENTAUR with a pair of the Navy's new jet fighter, the Hawker Seahawk. The Seahawks were assigned to No 806 Squadron which converted to the jets during March of 1953.

This Sea Fury FB 11 (WJ244) was assigned to the Fleet Requirement Unit (FRU) based at Hurn. One of the duties of an FRU was target towing for Naval gunnery practice. The starboard wing leading edge is rather weathered and a different color replacement aileron was installed.

A Sea Fury FB 11 of No 804 Squadron on final approach for landing aboard HMS GLORY. The Sea Fury was designed with large trailing edge flaps to allow the aircraft to make a fairly slow final landing approach.

Wing Fold

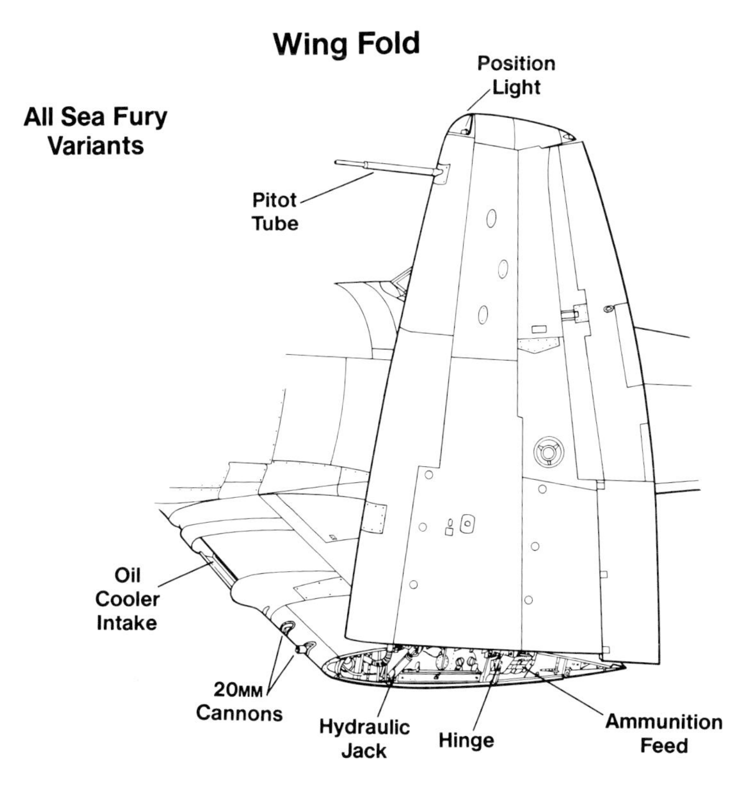

This Sea Fury has RATOG tubes mounted under the center of the fuselage. The absence of a tail hook suggests that the aircraft was assigned to the reserves for shore based use by an RNVR Squadron based at Culdrose. The colored spinner denotes that the aircraft was flown by the unit commander.

This Sea Fury has the propeller removed and has a wheel brace supporting the port wing. A tractor is preparing to take the FB 11 under tow by the tail hook. The reason for the wheel brace is unclear since the landing gear appears to be intact.

Rows of Sea Fury FB 11s repossessed by the Hawker company during 1957 for overhaul and export. The aircraft in the foreground have had their Naval camouflage removed and carry civil registrations. Some of these aircraft were later supplied to Cuba and Burma.

Sea Fury T Mk 20

Shortly after the end of the Second World War a Hurricane Mk IIC was modified to accommodate a second cockpit. Experience with this conversion led Hawker to commence work on a similar twin-seat version of the Sea Fury after interest was expressed by the Iraqi Government during early 1947. This aircraft was intended as one of four which the Iraqi student pilots could use as a transition trainer before going on the single seat Sea Fury. While construction of the prototype was taking place, the Royal Navy expressed a similar interest and intervened with a purchase order for the prototype (VX818) while offering to share the subsequent flight test data with the Iraqis.

The prototype made its initial flight at Langley on 15 January 1948. The pilots' positions were enclosed in separate cockpit canopys, but the aerodynamic loads from this layout were to prove potentially fatal when the rear canopy collapsed during a test flight. The incident forced Hawker to remove the rear cockpit windscreen, which was of no benefit to the instructor, as well as the rear section of the forward cockpit, the space between the cockpits being linked by a perspex "tunnel."

Other modifications included: reducing the armament to two Hispano Mk 5 cannons with the gun bays being used to house equipment normally carried in the fuselage, repositioning of the fuselage whip antenna to a location just forward of the front cockpit, deletion of the tail hook, installation of a non-retractable tailwheel and a periscope was installed above the rear cockpit gun sight to allow the instructor to aim weapons over the head of the student in the front cockpit. After successfully completing a series of test programs both by the company and A&AEE, the aircraft was ordered into production as the Sea Fury T 20.

A total of sixty-one Sea Fury T 20s were built and their principal use was as conversion/transition trainers, not just for front-line squadrons but for Fleet Air Arm Reserve units as well. The RNVR units were to take over Sea Fury FB 11s as they were relinquished by the fleet squadrons between 1951 and 1953. Royal Navy Volunteer Reserve (RNVR) pilots were known as "Weekend Fliers" because this was the time when they were generally available for training. Although many of them were experienced WW II pilots with Hellcat or Corsair experience, the Sea Fury T 20 was a positive factor in converting them to what was a very powerful and sometimes unforgiving naval fighter.

The total of T 20s produced was to prove ample for the Royal Navy's needs as well as those of the Iraqi and Pakistani Air Forces. The aircraft exported to these two services retained the separate cockpit configuration first installed on the prototype.

When the use of the T 20 was terminated during the mid-1950s, twenty were released for export. Burma purchased three (UB451-453) during 1957 and in 1957 Cuba negotiated for two aircraft. The remainder went to Germany between 1957 and 1962. These aircraft had the gun sight periscope pylons deleted and were modified with target towing equipment.

This T 20 trainer prototype with its twin cockpits and periscopic gunsight was overall Silver dope with Yellow fuselage and wing bands. The wing walkways were Black. This aircraft was originally to be built for Iraq but was taken over by the Royal Navy.

Canopy

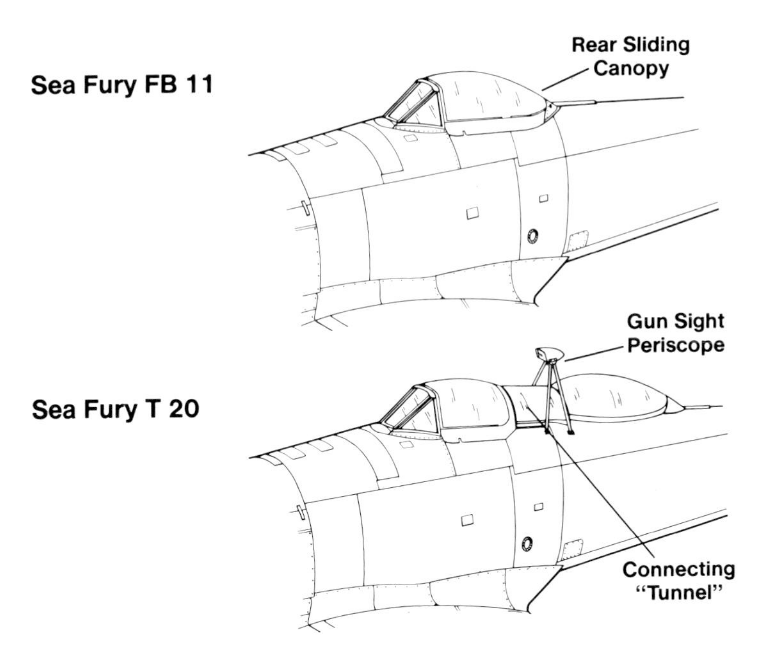

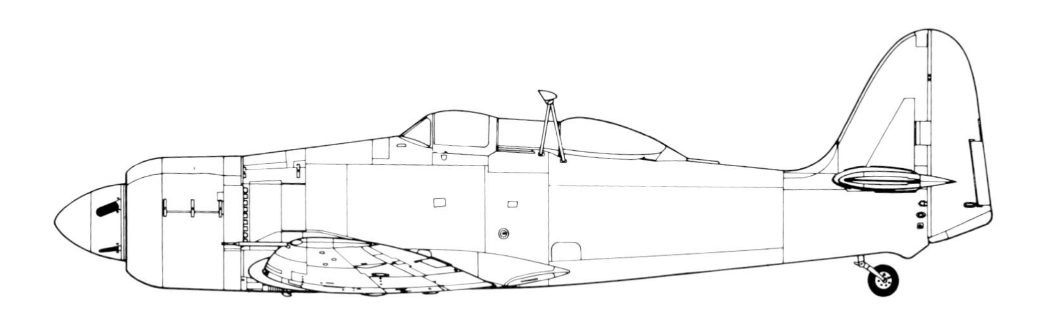

Specifications

Hawker Sea Fury T 20

Wingspan ....34 feet 3¾ inches
Length ....34 feet 8 inches
Height ....15 feet 10 ½ inches
Empty Weight .... 8,700 pounds
Maximum Weight ....11,930 pounds
Powerplant...................One 2,550 hp Bristol
Centaurus radial.

Armament....................Two 20MM cannon.
Two 500 or 1,000 pound
bombs and twelve rockets.

Performance
Maximum Speed ....445 mph
Service ceiling ....35,600 feet
Range.......................1,310 miles (with tanks)
Crew ....Two

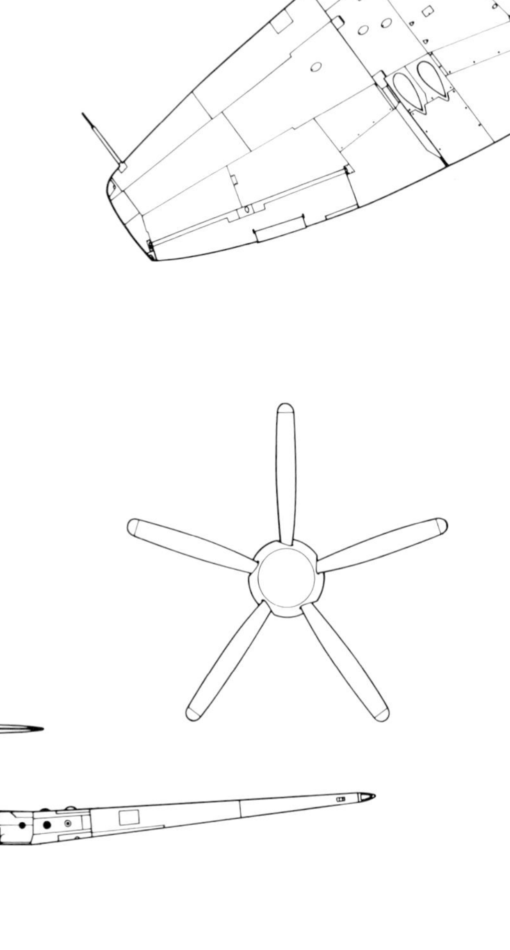

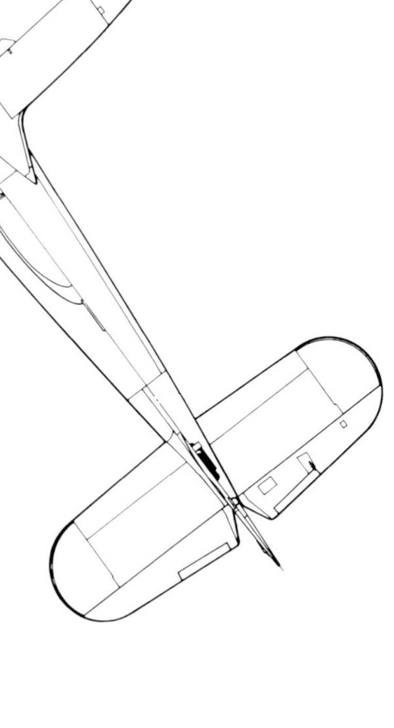

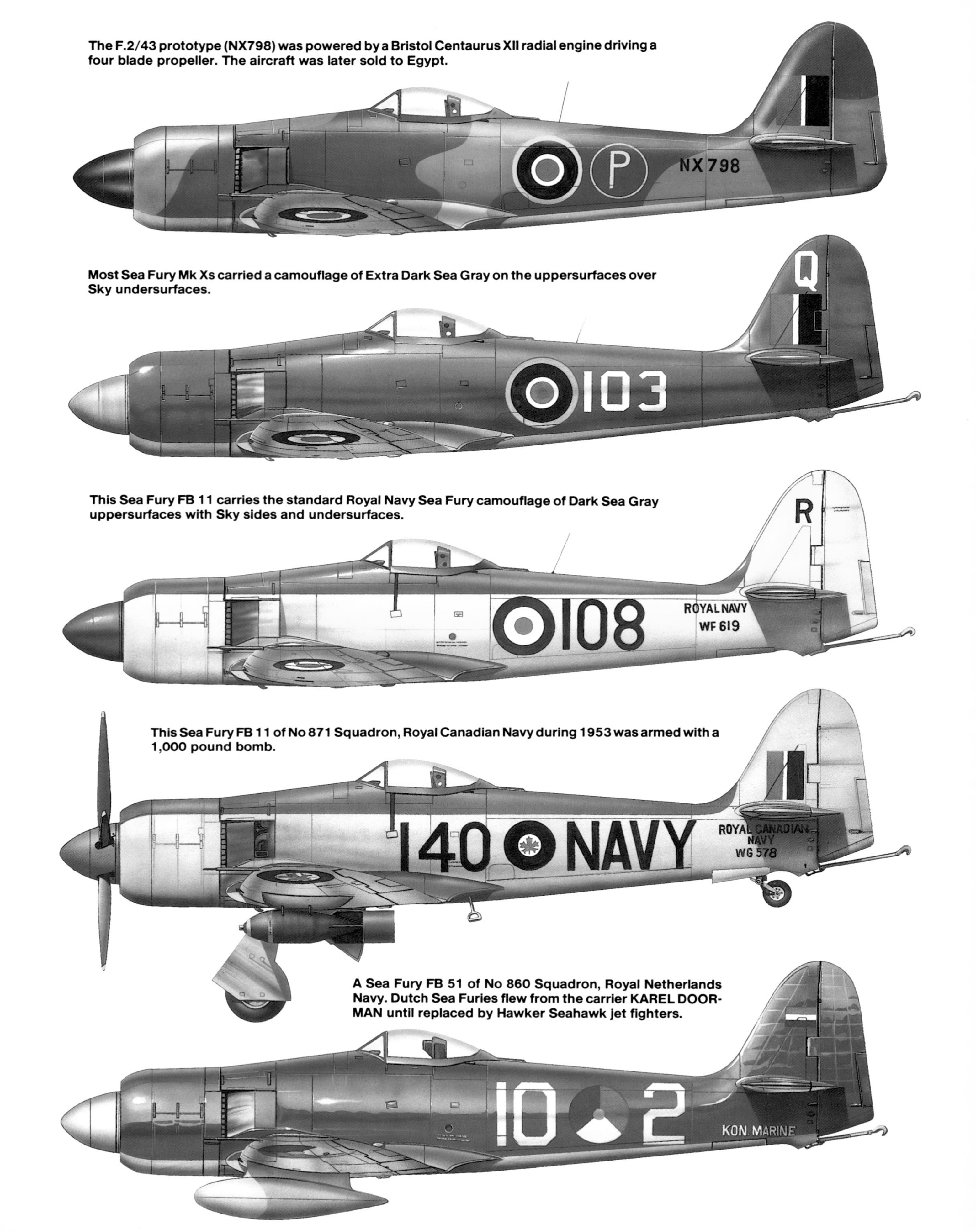

The F.2/43 prototype (NX798) was powered by a Bristol Centaurus XII radial engine driving a four blade propeller. The aircraft was later sold to Egypt.

Most Sea Fury Mk Xs carried a camouflage of Extra Dark Sea Gray on the uppersurfaces over Sky undersurfaces.

This Sea Fury FB 11 carries the standard Royal Navy Sea Fury camouflage of Dark Sea Gray uppersurfaces with Sky sides and undersurfaces.

This Sea Fury FB 11 of No 871 Squadron, Royal Canadian Navy during 1953 was armed with a 1,000 pound bomb.

A Sea Fury FB 51 of No 860 Squadron, Royal Netherlands Navy. Dutch Sea Furies flew from the carrier KAREL DOORMAN until replaced by Hawker Seahawk jet fighters.

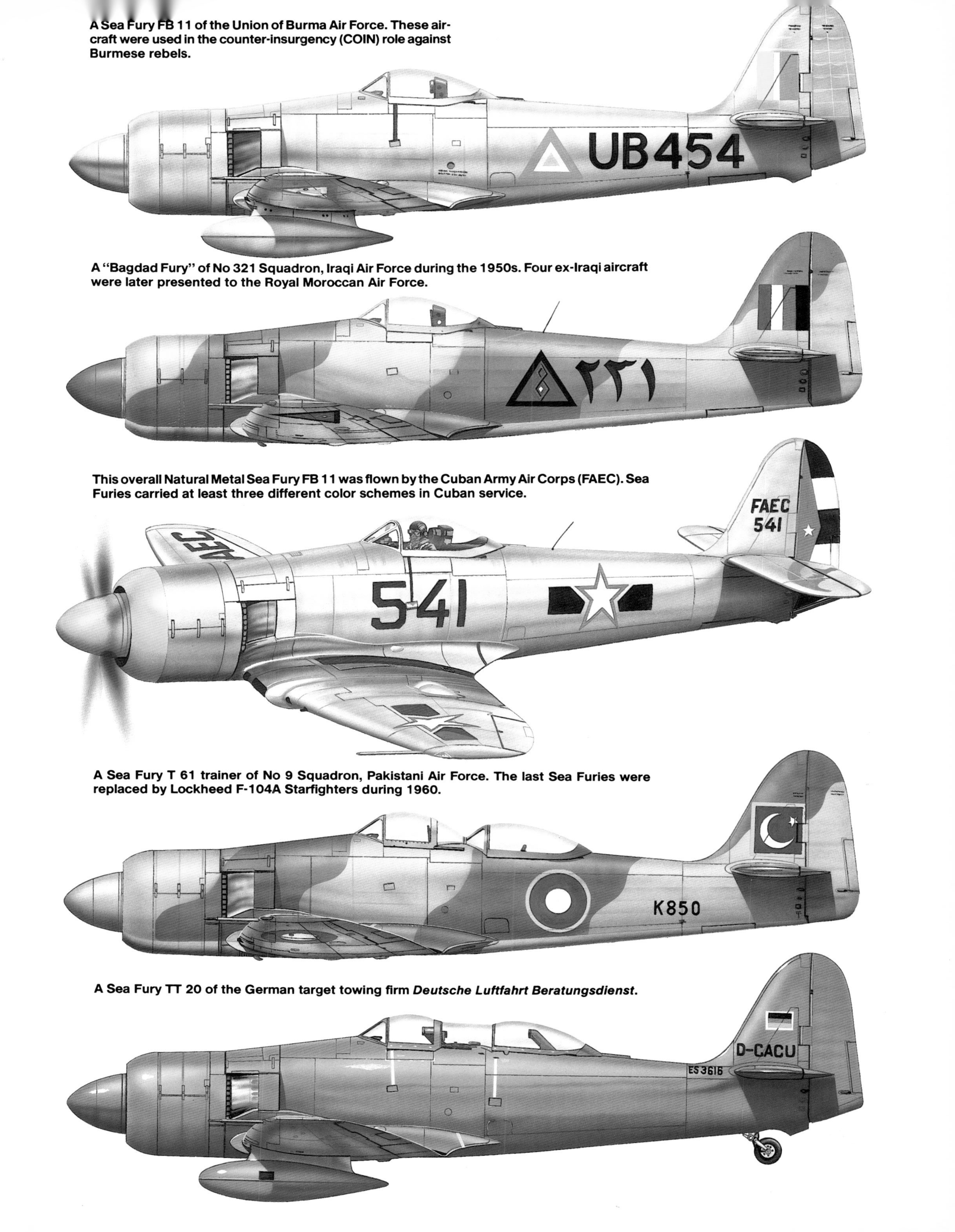

A Sea Fury FB 11 of the Union of Burma Air Force. These aircraft were used in the counter-insurgency (COIN) role against Burmese rebels.

A "Bagdad Fury" of No 321 Squadron, Iraqi Air Force during the 1950s. Four ex-Iraqi aircraft were later presented to the Royal Moroccan Air Force.

This overall Natural Metal Sea Fury FB 11 was flown by the Cuban Army Air Corps (FAEC). Sea Furies carried at least three different color schemes in Cuban service.

A Sea Fury T 61 trainer of No 9 Squadron, Pakistani Air Force. The last Sea Furies were replaced by Lockheed F-104A Starfighters during 1960.

A Sea Fury TT 20 of the German target towing firm *Deutsche Luftfahrt Beratungsdienst*.

This Sea Fury T 20 (VZ346) of No 1833 (RNVR) Squadron is on display during a family day. The aircraft has a Yellow fuselage band and carries the BR tail code which identifies the aircraft as being based at RNAS Bramcote.

This Sea Fury T 20 (VX301) is believed to be overall Dark Blue with a Red cowling and spinner and all lettering in White. It was the eighteenth aircraft of the first T 20 production batch.

This was the first production Sea Fury T 20 trainer. The aircraft was overall Silver and carried a Red and Yellow fuselage band. The T 20 had a fixed tail wheel and had the tail hook deleted. The aircraft in the background is a Vampire jet fighter.

Exports

Canada

Canada was one of two Commonwealth countries to make use of the Sea Fury FB 11. During 1947 two Sea Fury Mk X aircraft (TF901 and TF909) were sent to Canada. TF901 was used for demonstration and handling trials with the Royal Canadian Navy, while TF909 was sent to the Winter Experimental Flight at RCAF Station Edmonton, in Alberta for cold weather tests. After successfully completing the handling tests, the Sea Fury was selected to equip two Royal Canadian Navy squadrons with an initial order being placed for twenty-five aircraft.

The two units that were assigned Sea Furies were Nos 803 and 883 Squadrons. Originally, conversion training was to have been conducted in Northern Ireland at RNAS Eglinton but, in the event, only No 801 Squadron pilots were trained in Ireland. Initial deliveries for the Canadian order were delivered from Hawker to the Receipt and Dispatch Unit (RDU) at RNAS Culdrose. After clearance, they were loaded aboard HMCS MAGNIFICENT for transfer to Canada. Over the next four years a further fourteen aircraft were delivered.

The Canadian Sea Fury's were externally identical to Royal Navy Sea Fury FB 11s. When first delivered, the RCN aircraft were painted in Extra Dark Sea Gray uppersurfaces with Sky undersurfaces. Aircraft in this scheme were delivered to No 883 Squadron between 1948 and 1949. Canadian aircraft carried C 1 style fin flashes and Maple Leaf roundels on the wings and fuselage. Later the aircraft were repainted in Dark Sea Gray uppersurfaces with Sky undersurfaces and sides.

Until 1952/53 the Canadian Navy aircraft carried full squadron code letters. After that time the code letters were replaced by a number in the 100 range along with the word "NAVY" in Black on the fuselage. At this same time the camouflage was altered with Gloss Light Sea Gray replacing the Sky color.

During 1952 the squadrons' designations were changed with the prefix VF being added and the original squadron numbers being changed to 870 and 871 (VF-870 and VF-871).

Canadian Sea Furies were destined to spend the bulk of their operational careers ashore flying out of the Naval Shore Base HMCS SHEARWATER with their limited service at sea being on board HMCS MAGNIFICENT. During 1955, VF-870 gave up its Sea Furies in favor of the McDonnell F2H Banshee, being followed by VF-871 during 1956. The Sea Furies were placed in storage at Debert although three were transferred to the Canadian Civil Register (all of these ultimately being exported to the United States).

Four Sea Fury FB 11s of No 803 (RCN) Squadron fly a right echelon formation over the Canadian coast during 1948. The aircraft carry the Extra Dark Sea Gray over Sky camouflage but have later style fuselage roundels with the Maple Leaf emblem.

A pair of Royal Canadian Navy Sea Fury FB 11s (TF996 and TF999) of No 803 Squadron in flight near their Dartmouth base in Nova Scotia during 1950. Both aircraft carry the early camouflage with White fuselage codes.

A Royal Canadian Air Force DeHaviland Vampire jet fighter leads a Sea Fury FB 11 of No 806 (RN) Squadron (right) and a Sea Fury FB 11 of No 803 (RCN) Squadron (left). The aircraft were participating in the International Air Exhibition celebrating the Anniversary of New York.

Still bearing her original ROYAL NAVY logo, camouflage and markings, this Royal Canadian Navy Sea Fury FB 11 (VW552) taxies past other RCN aircraft in the early camouflage at Dartmouth on 2 June 1949. The aircraft was later repainted in full RCN markings.

LT Bell-Irving in the cockpit of his RCN Sea Fury FB 11 of No 871 Squadron on the snow covered ramp at Dartmouth, Nova Scotia, on 31 January 1952. The camouflage is Dark Sea Gray with Gloss Light Gray sides and undersides.

A trio of Sea Furies of No 871 Squadron flies over a solid cloud layer while operating from Halifax, Nova Scotia, on 26 May 1952. These aircraft carry full squadron codes and reduced size fuselage and wing roundels.

A 1,000 pound bomb is loaded on the wing pylon of this Sea Fury FB 11 (WG573) of No 871 Squadron at Rivers, Manitoba, on 4 January 1954. The underwing serial number has been replaced by the word NAVY in large Black letters.

This Royal Canadian Navy Sea Fury FB 11 was aboard the American aircraft carrier USS WASP during 1952. Unlike Royal Navy Sea Furies, Royal Canadian Navy aircraft carried fin flashes.

This Sea Fury FB 11 (TG115) runs up to full power while the pilot waits for the launch signal from the Flight Deck Officer on HMS MAGNIFICENT during June of 1953. The plane guard helicopter, an S55, has taken up station off the ship's starboard side. The nose art on the Sea Fury is a diving eagle.

Australia

The second Commonwealth country to operate the Sea Fury FB 11 was Australia. Two front line Navy units were re-established and equipped with Sea Furies. The first was No 805 Squadron which was reformed at Eglinton, Northern Ireland on 25 August 1948. Thirteen Sea Fury FB 11s were assigned and after aircrew training was completed the aircraft and crews were embarked in HMAS SYDNEY for the trip back home. The second squadron, No 808 Squadron, was similarly reformed at Saint Merryn, England on 25 April 1950, also receiving thirteen aircraft. The squadron embarked in SYDNEY during August of 1950 for transfer to Australia.

Both units would see combat aboard HMAS SYDNEY off Korea during late 1951. Similarly both Squadrons later served on board HMAS VENGEANCE (a Light Fleet Carrier on loan from the RN between 1952/55).

No 808 was first to retire its Sea Furies when it was disbanded at Nowra in October of 1954. In contrast, No 805 Squadron was to operate the Hawker fighter for another four years before it too was disbanded at Nowra.

Sea Furies were also used by No 724 (RAN) Squadron which was a Miscellaneous Air Squadron used as an Operational Training School and Aircrew Conversion Unit, based out of Nowra New South Wales. This unit had been formed during June of 1955 and although its aircraft were temporarily transferred to Nos 805 and 851 Squadrons between 1956 and 1961, it was to utilized the Sea Fury until October of 1962.

A third unit also operated the Sea Fury FB 11 within the RAN. This unit, No 850 Squadron, was formed at Nowra on 12 January 1953 with twelve aircraft. Before being disbanded in August of 1954, it had operated briefly aboard HMAS SYDNEY off Korea just after the end of the Korean War.

Three Sea Furies overfly their carrier, HMAS SYDNEY. These aircraft carry the early and standard Sea fury camouflage, indicating that the cruise took place prior to the deployment of the aircraft and ship to Korea during 1951.

A section of Royal Australian Navy Sea Fury FB 11s escorts a Hastings transport carrying the First Sea Lord, Louis Montbatten, on his 1956 Australian tour. The standard Sea Fury camouflage was later changed to overall Dark Blue with White lettering.

A Royal Australian Navy FB 11 is hoisted aboard HMAS SYDNEY in King George Dock, Glasgow, Scotland, during late 1950. The ship was picking up the Sea Furies and Fireflies ordered by Australia. The aircraft's rudder and elevators were secured to prevent damage while being transported.

A Sea Fury FB 11 of No 808 (RAN) Squadron carries the "K" tail code identifying it as being based aboard HMAS SYDNEY. The number "01" on the landing gear door indicates that this Sea Fury was flown by the unit squadron commander.

An overall Dark Blue Sea Fury FB 11 of No 724 Squadron, an Operational Training School which used a variety of aircraft types. The Australians were the only country to adopt this color scheme. The aircraft carries the later Red "Wallaby" insignia and the spinner and code letters are in White.

This Sea Fury FB 11 (WH587) was one of the last four Sea Furies supplied to the Royal Australian Navy by Hawker. The aircraft was assigned to No 808 Squadron.

Netherlands

The Netherlands had possessed a strong pre-War Navy and with large colonies in the Netherlands East Indies (today known as Indonesia) still under its control during 1945, it was logical for its Naval forces to be re-built as quickly as possible for deployment to the Far East. As part of this rebuilding effort, carrier aircraft were deemed a necessary component of the Navy's overall strength. As a result, it was not surprising that a sales drive by Hawker during 1946 resulted in a response by a Dutch purchasing commission looking for suitable carrier based aircraft.

The Dutch placed an initial order for ten Sea Fury FB 11s on 21 October 1946 under the export designation Sea Fury FB 50. These aircraft were intended to serve on board the ex-Royal Navy escort carrier HrMs NAIRANA. Some four years later, a second order for fourteen aircraft was placed with Hawker to supplement the Sea Furies being built by Fokker. Twenty-four license-built aircraft were completed by the Dutch Fokker company bringing overall deliveries of Sea Fury FB 50s to the Dutch Navy to forty-eight aircraft.

The Sea Furies were flown by the Fighter Pilot Combat School at Valkenburn Naval Air Station. This unit was later redesignated as No 860 Squadron. The unit received its initial carrier training alongside Royal Navy Sea Fury squadrons aboard HMS INDOMITABLE and HMS ILLUSTRIOUS during 1951/52. On 24 September 1952, the squadron was declared fully operational and flew off the NAIRANA and the light carrier HrMs KAREL DOORMAN which replaced the NAIRANA. No 860 Squadron continued to fly the Sea Fury until 15 June 1956, when it transferred its remaining aircraft to No 3 Squadron.

The Sea Fury FB 50 formed the fighter element aboard the KAREL DOORMAN from 1952 until late 1959, being replaced by another Hawker fighter aircraft, the Seahawk.

The second Sea Fury produced for the Royal Netherlands Navy on the grass at the Hawker facility prior to delivery. A total of twenty-four Sea Fury Mk 50s were ordered from Hawker with a further twenty-four being built under license in Holland by Fokker.

A lineup of Dutch Navy Sea Fury Mk 50s of No 860 Squadron visits England for a joint Dutch Navy/Royal Navy training exercise during 1951. These aircraft were all Hawker-built Mk 50s.

The pilot of this Royal Netherlands Navy Sea Fury Mk 50, dives off the port wing through the fire from the ruptured fuel tank after Sea Fury number 27 crashed on board the Dutch carrier KAREL DOORMAN.

This Sea Fury on the KAREL DOORMAN engaged the crash barrier and nosed over. The engine was torn off the Sea Fury and fuel from the ruptured fuel lines has caught fire. Deck crewmen are already moving toward the aircraft with fire fighting equipment.

This Sea Fury was a license-built Fokker Mk 50 as indicated by the aircraft number. Aircraft numbers 1 through 24 were carried by Hawker-built aircraft and higher numbers were assigned to Fokker-built aircraft. The aircraft was visiting RAF Acklington on 20 September 1952.

Pakistan

Hawker fighters were flown by the Pakistani Air Force since that nation first achieved its independence. The first Hawker fighters were Tempest IIs which were used to form Nos 5 and 9 (PAF) Squadrons during August of 1947.

During 1949, an initial order was placed for fifty Sea Fury FB 11s under the export designation Sea Fury FB 60 (Serials L900 through L949) with deliveries scheduled to begin the following year (1950). A grand total of eighty-seven new built Sea Fury FB 60s were purchased between 1950 and 1952. Included in the contract was the second Fury F2/43 prototype (NX802) which was completely refurbished and brought up to FB 60 standards. The FB 60 was basically a standard Sea Fury FB 11, with the naval equipment (tail hook and catapult hook) deleted.

Between 1953 and 1954 an additional five ex-Fleet Air Arm aircraft (refurbished by Hawker) were purchased, along with five two-seat trainers designated the Sea Fury T 61. These aircraft differed from the Royal Navy T 20 Sea Fury trainer in that they had separate canopies and windscreens for each cockpit, an arrangement the Royal Navy had abandoned after an accident with the T 20 prototype. Four of the Sea Fury T 61s were new construction aircraft and the fifth was released from storage where it had been placed after being declared surplus to an Iraqi Air Force order.

A total of ninety-three Sea Fury Mk 60s were purchased by the Pakistani Air Force. The presence of the tail hook fairing under the fuselage indicates that L976 was one of the five ex-FAA aircraft supplied during 1953/54.

This Sea Fury Mk 60 of the Pakistani Air Force attempted to set new speed over distance records during its delivery flight. Two records were attempted: the London-Rome record and the London-Karachi record.

The Pakistani Air Force assigned the Sea Fury FB 60 to three front line units, Nos 5, 9 and 14 Squadrons. Additionally, No. 2 Fighter Conversion School also used the Sea Fury T 61 to train pilots for the operational units. The aircraft was replaced during 1955 in Nos 5 and 14 Squadrons by the North American F-86F Sabre, but it was another five years (1960) before No 9 Squadron converted from the Sea Fury to Lockheed F-104A Starfighter.

The Sea Fury T Mk 61 differed from the T 20 in having separate windscreen and canopies for each cockpit. This T 61 (K850) of the Pakistani Air Force was originally built for the Iraqi Air Force but became the first of five delivered to Pakistan during 1949/50.

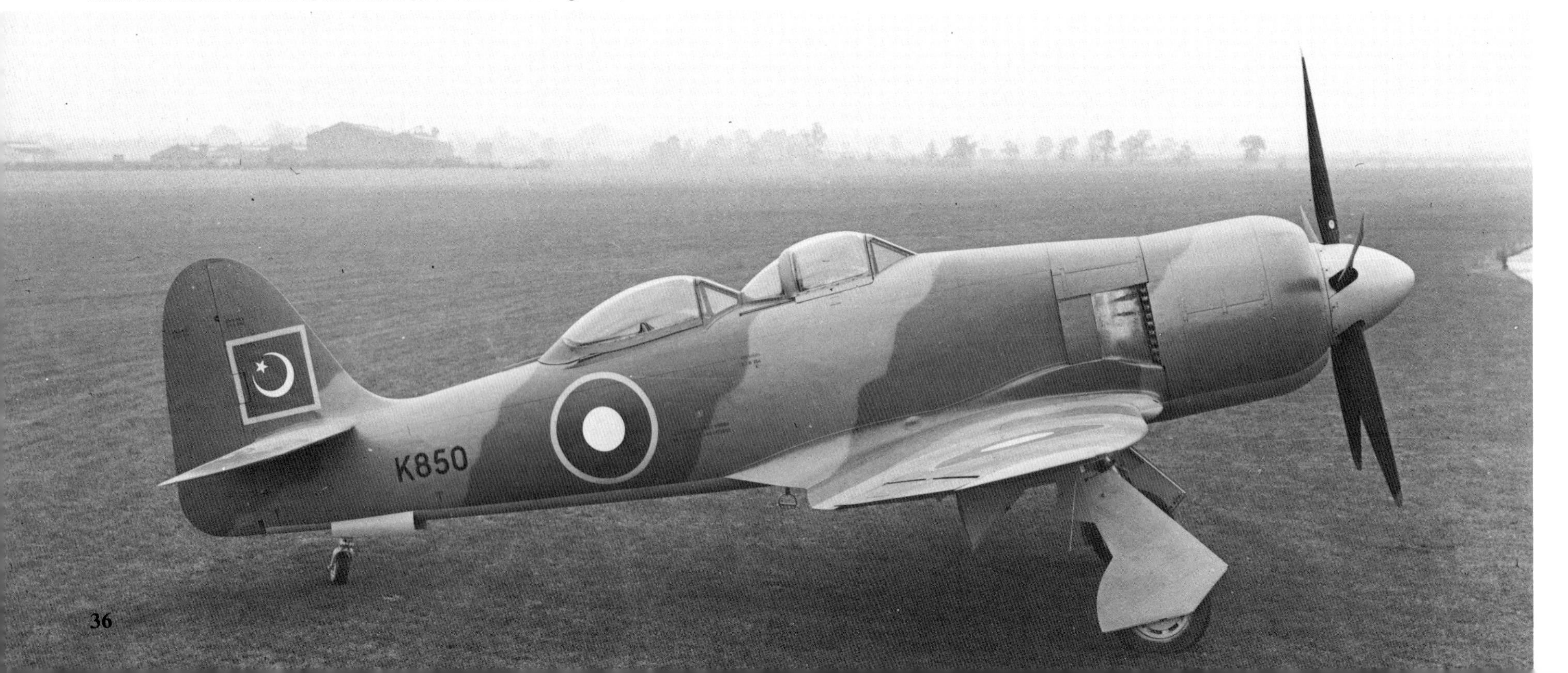

Two Seat Trainers

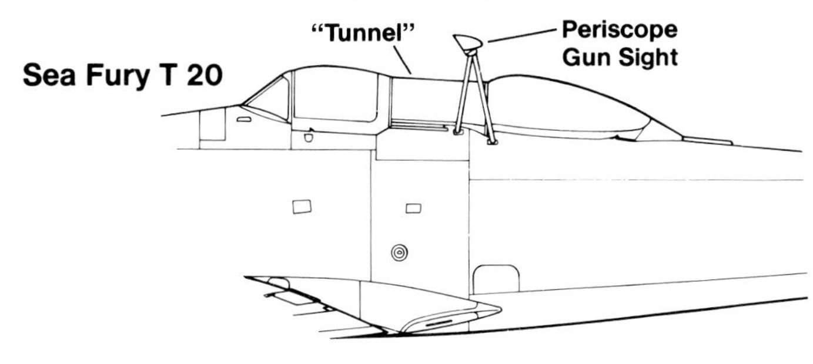

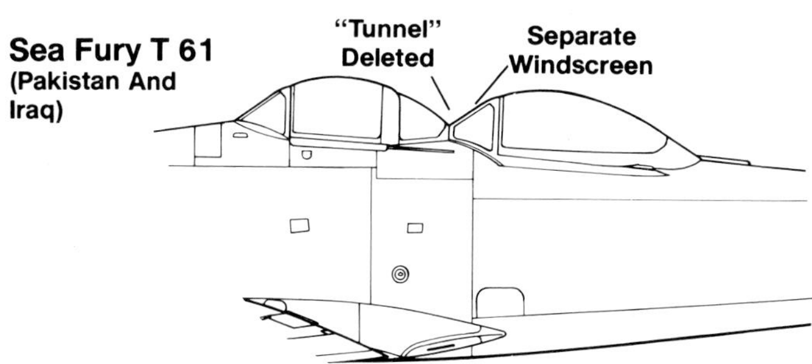

Iraq

During December of 1946, the Iraqi Air Force placed orders for a total of thirty Sea Fury FB 11s plus four two seat trainers. Delivery took place during 1947/48 with the trainer section of the order being reduced from four down to a pair of trainers. In July of 1951, a further twenty-five fighters and three trainers were ordered, although because of Royal Navy needs to meet operations in Korea, deliveries against this order did not start until May of 1952.

All Iraqi Sea Furies were de-navalised with the tail hook, catapult launch hook and other naval equipment deleted. The tail hook fairing was deleted and the rudder extended and rounded off. Otherwise they were standard Sea Fury FB 11s. The aircraft were referred to in Hawker documents as "Bagdad Furies."

The Sea Fury was flown by Nos 1, 4 and 7 Squadrons of the Iraqi Air Force until the early 1960s when they were replaced by Hawker Hunter jet fighter-bombers.

Recently, information has come to light that four ex-Iraqi Air Force Sea Furies were transferred to Morocco shortly after that country achieved independence. These aircraft were reportedly flown for a short time, being replaced by Soviet MiG-17 Fresco fighters.

During the late 1970, a number of the Iraqi Sea Furies that had been in open storage outside Bagdad were purchased by David Tallichet and Ed Jurist and imported to the United States. During 1980, one of these aircraft completed restoration (in Royal Canadian Navy markings) and was registered N19SF.

One of the fifty-five Sea Fury Mk 60s supplied to the Iraqi Air Force. Iraqi Sea Furies had the tail hook and fairing deleted. The aircraft in the background appears to be the prototype T 20 trainer.

One of the three Iraqi Air Force T 61s on the ramp at Habbaniya Air Base. The T 61s were operated by No 1 (IAF) Squadron in the conversion training role. Unlike Royal Navy T 20 aircraft, these retained the original twin canopies first tested on the prototype.

Egypt

Along with the original F2/43 prototype (NX798), Egypt purchased a further twelve FB 11s which were delivered during 1950/51. A number of these aircraft were still in Egyptian service at the time of the 1956 Suez Crisis, although none were noted as having been met in combat by either the Israelis or the Anglo/French forces. On the other hand, at least one Sea Fury was destroyed on the ground during a strafing attack by Anglo-French forces.

Burma

Between 1957/58 Burma received some twenty-one aircraft in two separate groups. Eighteen ex-Fleet Air Arm Sea Fury FB 11s formed the first delivery, of which three were converted for target towing duties and designated as Sea Fury TT 11s. These aircraft retained their tail hooks. The second delivery was made up of three refurbished two seat trainers. The Sea Fury was used for the internal security and in the counter-insurgency role in Burmese service, being replaced during 1968 by Lockheed AT-33 Shooting Star armed jet trainers.

The first of eighteen Sea Fury FB 11s destined for the Union of Burma Air Force. The aircraft were all re-furbished ex-Fleet Air Arm aircraft. Burmese national markings were Blue/White and Yellow and the serial was in Black.

Cuba

During 1957, the Batista government of Cuba ordered a total of seventeen Sea Furies, fifteen single seat FB 11s and two Sea Fury T 20 two seat trainers, all being ex-Fleet Air Arm aircraft refurbished by Hawker. Despite published reports stating that the Sea Furies were not in use prior to the Castro overthrow of the Batista government, it is known that they were actually delivered during 1958, assembled and flown in Cuban Army Air Corps (FAEC) markings, although there are no reports that the Sea Furies saw combat during the revolt.

After the overthrow of Batista, the Castro government experienced numerous problems with the Sea Furies, mainly due to the fact that most of the qualified maintenance personnel and pilots had been purged from the air force. It is known, for example, that there were problems with starter units and these were replaced with starters taken from surplus P-47D Thunderbolts.

The Sea Furies did, however, engage in combat operations during the Bay of Pigs invasion of 1961. At the time of the invasion, there were at least five operational Sea Furies in Cuba. Of these, one was reportedly destroyed on the ground during the opening raids by Liberation Air Force B-26Bs.

During the seaborne landings, one Sea fury, flown by MAJ Carrera-Rolas, attacked the Brigade ship HOUSTON with rocket fire. The rockets damaged the HOUSTON and she was run aground to keep from sinking. The Sea Fury was damaged by a patrolling Brigade B-26, but was able to make a successful landing at San Antonio de Los Banos Air Base. Another Sea Fury, flown by Gustavo Bourzac, was attempting to strafe another ship when he was attacked by two Brigade B-26s. Again thanks to the Sea Fury's superior speed, he was able to escape undamaged.

In another engagement, LT Douglas Rudd was attacking a Brigade B-26 when his Sea Fury picked up an unexpected wingman. As if from nowhere, an unmarked Douglas A-4 Skyhawk had joined on his port wing. Faced with the A-4's superior performance and not knowing the pilot's intentions, Rudd broke off his attack and returned to base. The A-4 had been flown by CDR Jim Forgy from the USS ESSEX and, unknown to Rudd, he had just been denied permission to fire on the Sea Fury. Another Brigade B-26, Puma One, reported that he was saved from a Sea Fury by unmarked A-4s that positioned themselves between the attacking Sea Fury and the B-26.

A Brigade C-46, dropping supplies over the invasion beach, was jumped by a Sea Fury; however, the Commando pilot managed to shake the attacking Sea Fury by dropping to wave height and slowing the C-46 to just above stall speed. Unable to stay behind the C-46, the Sea Fury , flown by Carlos Ulloa, make several passes before he stalled and crashed into the sea — the only Sea Fury lost in air combat during the invasion.

At least three different color schemes have been noted on Cuban Sea Furies. The aircraft were flown in overall natural metal with FAEC markings and Black numbers and codes. Another reported scheme was overall Light Gray with FAEC markings and Black numbers and codes. The FAEC aircraft had colored cowling rings and spinners to denote their flight assignment, with the natural metal aircraft having Red, Yellow, or White spinners and cowl rings. The overall Gray aircraft had either Red or Yellow cowl rings and spinners.

After the Sea Furies were absorbed into Castro's revolutionary air force (FAR) the aircraft were repainted in overall Dark Green with FAR markings and numbers in Yellow. The old FAEC star and bar insignia was retained, although after a short period, the rudder stripes were reported deleted. There are at least two Sea Furies remaining in Cuba, both on public display. Neither aircraft carries an accurate operational paint scheme.

This overall Dark Green Cuban Sea Fury was flown by the Revolutionary Air Force after Castro overthrew the Batista government. The lettering FAR on the tail is believed to have been in Yellow. Later the Cuban star and bar insignia was applied to the fuselage and wings.

FAR 42 was previously FAEC 542 in the Cuban Army Air Corps under Batista. The aircraft had been recently repainted in overall Dark Green. Later the fuselage national insignia was added and, for a short time, rudder stripes were carried. The serial under the wing is believed to have been in Yellow. Reportedly, the Sea Furies carried this camouflage at the time of the Bay of Pigs landings.

Still later, (December of 1976) FAR 541 was again repainted with the current style of national insignia. This insignia was never carried by the Sea Furies during their operational carreers in Cuba.

After the Bay of Pigs landings, the Sea Furies were retired and FAR 41 was repainted in a striped camouflage pattern for display purposes. The aircraft never carried this type of camouflage in operational service.

FAR 541 has carried a number of colorful, if not totally accurate, camouflage patterns while on public display. Currently, the aircraft carries this two tone camouflage with late Cuban style national insignia and an unusual style of lettering.

West Germany

The final Sea Fury order placed by an overseas source was almost totally for Sea Fury T 20s. During the late 1950s, the re-emerging German Luftwaffe required, among other things, target towing services and aircraft. As a result, ten T 20s, that had been previously de-navalised by the Ministry of Supply, were purchased by a German company, *Deutsche Luftfahrt Beratungsdienst*, who were under contract to provide target tow services for the Luftwaffe. The aircraft were delivered between 1958 and 1960.

The target towing winch gear was of Swiss design and was tested by Hawker on two of the aircraft at the Hawker plant. The remaining eight aircraft were modified by DLB after they were delivered from modification kits. After installation of the target towing equipment the aircraft were designated as Sea Fury TT 20.

Seven additional aircraft were ordered during 1962; six from Hawker and the seventh (an FB 11) from Holland. These aircraft remained in service well into the 1970s, operating out of three German airfields. Five aircraft were written off in accidents and one was retained for a museum display after they were retired from active service. The remainder were sold to private sources, mainly in Britain or the United States.

Target Tow Winch

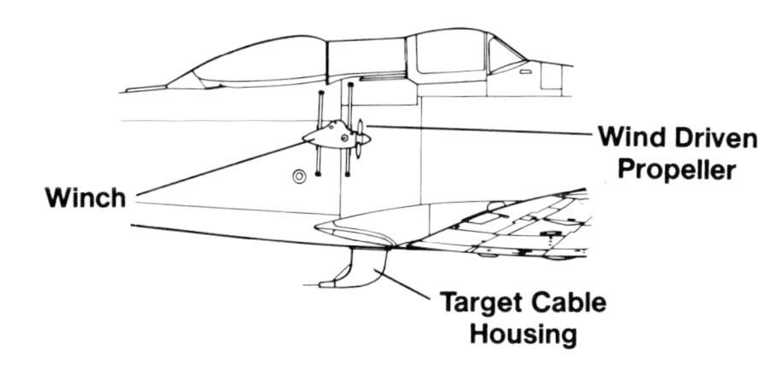

This is one of the two Sea Fury TT 20s modified by Hawker for DLB. The Swiss target towing winch gear was tested on this aircraft before it was delivered to the German company. Later the aircraft was repainted overall Red for high visibility.

The target towing winch gear fitted to the West German Sea Fury TT 20 was of Swiss manufacture. This aircraft was one of two airframes modified and tested by Hawker prior to delivery to Germany. The remaining aircraft were modified in Germany.

West Germany purchased at least ten T 20s for target-towing work during the late 1950s and early 1960s. The aircraft were modified with the guns, gun sight periscope and rear cockpit controls deleted. A target-towing unit was mounted on the starboard fuselage side.

A *Deutsch Luftfahrt Beratungsdienst* (DLB) Sea Fury TT 20 on a snow covered ramp at its home base in Germany. The aircraft were used for a number of years after which they were sold on the private market in England and the United States.

Royal Navy Historic Flight

During 1989/90 the Royal Navy Historic Flight suffered the loss of both its flyable Sea Furies. During 1989 the sole flyable single FB 11 was lost when it had an unsafe landing gear indication and the pilot, John Beattie, was directed to bail out rather than try a landing. The second Sea Fury, a T 20, was lost on 14 July 1990, when the aircraft crashed while attempting a forced landing after suffering an engine failure on takeoff. The aircraft broke in half after striking some trees.

Civil Sea Furies

The Sea Fury has a long history of air racing in the United States, a history that started during 1962. Brian Baird, a retired RCAF pilot, purchased several ex-Canadian Navy Sea Furies and sold one to an American in Louisiana. This aircraft was to pass through several hands before being restored by Frank Sanders during 1969/70. The aircraft was painted in Royal Navy Korean War markings and was widely used in racing and in air shows.

Another Sea Fury that was widely known in the United States was registered N878M, belonging to the late Mike Carroll.This Sea Fury was highly modified with clipped wings, a small tear-drop canopy, installation of a "wet" wing, and a weight reduction of some 170 pounds (removal of wing fold gear, armor, and tail hook). With its "hot rod" color scheme, the Sea Fury was flown in several cross country races. After Carroll's death, the Sea Fury was purchased by Dr. Sherman Cooper and he flew it to first place in the USA Cup race held at San Diego during 1971. The aircraft later crashed at Mojave, California, during the California 1000 race.

Sea Furies have been flown extensively at the National Air Races in Reno, Nevada. Although not well suited to the short pylon style racing at Reno, Sea Furies have placed in a number of races. In an effort to make the Sea Fury more competitive, several conversions have been undertaken aimed at increasing the aircraft's speed. A number of 3,800 hp Pratt & Whitney R4360 twenty-eight cylinder engines were purchased by Frank Sanders, who installed one in an ex-Burmese Air Force T 20 trainer during 1982. A second conversion of a single seat Sea Fury was undertaken by Llyod Hamilton. The Sanders conversion was finished first and the aircraft, named Dreadnought, made its first flight on 6 August 1983. Dreadnought, flown by General Dynamics test pilot Niel Anderson, later won the 1983 Reno race at an average speed of 425.24 mph.

Part of the conversion was the replacement of the five blade Rotol propeller with a four blade propeller from a Douglas A-1 Skyraider. Additionally, to improve directional stability, the fin was raised some twelve inches, while the rudder was shortened some six inches to improve rudder pedal loads.

Another Sea Fury racing conversion is Lloyd Hamilton's Sea fury known as "Furious." This is another R4360 powered aircraft with much of the conversion parts being supplied by Frank Sanders. "Cottonmouth" is a re-engined T 20 registered in Canada but flown extensively in the United States. The aircraft was purchased by Jerry Janes and shortly after taking delivery the engine suffered a complete failure. As a result, Janes decided to re-engine the aircraft with a 3,700 hp Wright R3350 radial. During the conversion the

Sea Fury TF956 was originally restored for inclusion in a Hawker design display but was given to the Royal Navy Historic Flight painted in the colors and markings for an aircraft off HMS THESEUS during Korea.

original five blade propeller was replaced with an A-1 propeller and the rear seat controls were re-installed. The T 20 had been one of the aircraft sold to Germany as a target tow and the original rear seat controls had been deleted at that time. In its highly polished natural metal finish, "Cottonmouth" is a popular participant on the U.S. air show circuit.

During January of 1990 another Wright R-3350 conversion was registered by Howard Pardue. His ex-Iraqi Sea Fury carries a four blade propeller and has a two place cockpit covered by a enlarged single seat style canopy. The aircraft is painted in an RAF style camouflage. There are thirty-four Sea Furies on the FAA Register, five FB 11s and twenty-nine T 20s. This availability of Sea Furies and the availability of replacement engines should see additional racing Sea Furies joining the Reno races and the air show circuit in the future.

The Royal Navy Historical Flight's (RNHF) Sea Fury FB 11 flies alongside the British Concorde 002 during the Concord's arrival at RNAS Yeovilton for an Air Display on 4 March 1976.

The RNHF's Sea Fury was lost during 1989 when the pilot received an unsafe landing gear indication. He was directed to bail out rather than attempt a landing. The aircraft crashed into the sea and was lost.

This former Royal Australian Navy FB 11 was sold to Fawcett Aviation during 1963. It originally carried a Red color scheme with White trim and spinner. Later it was sold to Lloyd Hamilton who repainted the aircraft Green with Yellow trim and entered the aircraft in the Reno Race of 1972.

Spencer Flack owned this Sea Fury FB 11. The aircraft taxies past a collection of modern USAF aircraft during the Open Day at RAF Mildenhall during 1981. The Sea Fury was overall Red with a White spinner star and fuselage lines. The Sea Fury was subsequently destroyed while making a forced landing.

This highly colorful and highly modified Sea Fury was owned by Mike Carroll who rebuilt the aircraft for air racing. The canopy was replaced by a very small bubble, all excess weight was removed and the aircraft was given this wild Yellow and Red racing color scheme.

On the ramp at San Diego during 1971, Mike Carroll's Sea Fury reveals its clipped wings. Mike Carroll reduced the wing span of the Sea Fury to thirty-two feet from thirty-eight feet. Additionally, the aircraft had additional fuel tanks in the wings for long distance racing.

N232 is an ex-Royal Canadian Navy Sea Fury FB 11 owned and flown by Frank Sanders. The aircraft, on the ramp at San Diego alongside Mike Carroll's racing Sea Fury, was in San Diego to promote the USA 1000 air race held near San Diego during 1971.

One of the more attractive reconstructed Sea Furies was Frank Sanders' Sea Fury FB 11. The aircraft had been owned by Brian Baird and had been crashed by the late Bill Fornoff. Sanders rebuilt the Sea Fury and it is a regular on the air show circuit.

Although entered in several Reno races, N232 never won the event. The Sea Fury with its Centaurus engine proved ill suited for the short pylon style of racing at Reno. Over longer distances, the Sea Furies did well, winning several 1,000 mile races.

Sanders' Sea Fury FB 11 carries a Royal Navy Korean war color scheme, complete with invasion stripes. The aircraft was intended to represent LT Carmichael's MiG killer; however, the side number was mistakenly changed to "232" during painting. It should have been "114."

Frank Sanders' Sea Fury has since been repainted and no longer carries the Royal Navy colors. N2332 now has a Royal Canadian Navy scheme. This is fitting, since the aircraft was originally TG114 of the Royal Canadian Navy.

The Centaurus engine in Dreadnaught was replaced by a 3,800 hp Pratt and Whitney R-4360 Wasp Major in what has become known as a "Corncob" conversion. The Sea Fury T 20 formerly served with the Royal Navy and the Burmese Air Force. The fin has been lengthened and the rudder shortened to improve performance.

A highly polished "Cottonmouth" sits on the grass at Breckenridge, Texas, during 1989 as part of Howard Pardue's annual gathering of warbirds. Howard also owns and operates a R-3350 Sea Fury conversion.

"Cottonmouth" was converted to use a 3,700 hp Wright R-3350 engine after its Centaurus suffered a complete engine failure. The aircraft was one of the German target tows that was completely rebuilt with the new engine, four blade propeller and with the rear seat controls reinstalled.

RAF Aircraft from squadron/signal

1039

WELLINGTON
in action
Aircraft Number 76
squadron/signal publications, Inc.

1076

BRISTOL BLENHEIM in action
Aircraft Number 88
squadron/signal publications, inc.

1088

1096

1102

1111

squadron/signal publications